Getting Started with R

Getting Started with R

An Introduction for Biologists

Second Edition

ANDREW P. BECKERMAN
DYLAN Z. CHILDS

Department of Animal and Plant Sciences,
University of Sheffield

OWEN L. PETCHEY

Department of Evolutionary Biology
and Environmental Studies,
University of Zurich

OXFORD
UNIVERSITY PRESS

OXFORD
UNIVERSITY PRESS

Great Clarendon Street, Oxford, OX2 6DP,
United Kingdom

Oxford University Press is a department of the University of Oxford.
It furthers the University's objective of excellence in research, scholarship,
and education by publishing worldwide. Oxford is a registered trade mark of
Oxford University Press in the UK and in certain other countries

© Andrew Beckerman, Dylan Childs, & Owen Petchey 2017

The moral rights of the authors have been asserted

First Edition published in 2012
Second Edition published in 2017

Published in the United States of America by Oxford University Press
198 Madison Avenue, New York, NY 10016, United States of America

British Library Cataloguing in Publication Data

Data available

Library of Congress Control Number: 2016946804

ISBN 978-0-19-878783-9 (hbk.)
ISBN 978-0-19-878784-6 (pbk.)

DOI 10.1093/oso/9780198787839.001.0001

Printed and bound by
CPI Group (UK) Ltd, Croydon, CR04YY

Contents

Preface

Introduction to the second edition

This is a book about how to use R, an open source programming language and environment for statistics. It is not a book about statistics per se, but a book about getting started using R. It is a book that we hope will teach you how using R can make your life (research career) easier.

Several years ago we published the first edition of this book, aiming to help people move from 'hearing about R' to 'using R'. We had realized that there were lots of books about exploring data and doing statistics with R, but none specifically designed for people that didn't have a lot of experience or confidence in using much more than a spreadsheet, people that didn't have a lot of time, and people that appreciated an engaging and sometimes humorous initial journey into R. The first edition was also designed for people who did know statistics and other packages, but wanted a quick 'getting started' guide, because, well, it is hard to get started with R in some ways. Overall, we aimed to make the somewhat steep learning curve more of a walk in the park.

Over the past five years much has changed. Most significantly, R has evolved as a platform for doing data analysis, for managing data, and for producing figures. Other things have not changed. People still seem to need and appreciate help in navigating the process of getting started working with R. Thus, this new version of the book does two things. It retains our focus on helping you get started using R. We love doing this and we've been teaching this for 15 years. Not surprisingly, many of you are also finding that this getting-started book is great for undergraduate and graduate teaching. We thank you all for your feedback!

Second, we have substantially revised how we use, and thus suggest you use, R. Our changes and suggestions take advantage of some new and very cool, efficient, and straightforward tools. We think these changes will help you focus even more on your data and questions. This is good.

If you compare this second edition with the first, you will find several differences. We no longer rely on base R tools and graphics for data manipulation and figure making, instead introducing *dplyr* and **ggplot2**. We've also expanded the set of basic statistics we introduce to you, including new examples of a simple regression and a one-way and a two-way ANOVA, in addition to the old ANCOVA example. Third, we provide an entire new chapter on the generalized linear model. Oh, yes, and we have added an author, Dylan.

WHAT'S SO DIFFERENT FROM THE FIRST EDITION?

We teach a particular workflow for quantitative problem solving: have a clear question, get the right data for that question, inspect and visualize the data, use the visualization to reveal the answer to the question, make a statistical model that reflects your question, check the assumptions of the model, interpret the model to confirm or refute your answer, and clearly and beautifully communicate your answer in a figure.

In R there are many different tools, and combinations of these tools, for accomplishing this workflow. In the first edition of this book we introduced a set of 'classic' R tools drawn from the base R installation. These classic tools worked and, importantly, continue to work very well. We taught them in our courses for years. We used them in our research for years. We still use them sometimes. And as you start to use R, and interact with people using R, and perhaps share code, you will find many people using these classic tools and methods.

But the tools and their syntax were designed a long time ago. Many employ a rather idiosyncratic set of symbols and syntax to accomplish tasks. For example, square brackets are used for selecting parts of datasets, and dollar signs for referring to particular variables. Sometimes different tools that perform similar tasks work in very different ways. This makes for rather idiosyncratic instructions that are not so easy for people to read or to remember how to write.

So after much deliberation, and some good experiences, we decided that in this second edition we would introduce a popular and new set of tools contributed by Sir[1] Hadley Wickham and many key collaborators (`http://had.co.nz`). These new tools introduce a set of quite standardized and coherent syntax and exist in a set of add-on packages—you will learn exactly what these are and how to use them later. And you will also learn some base R. In fact, you will learn a great deal of base R.

We decided to teach this new way of using R because:

- The tools use a more 'natural language' that is easier for humans to work with.
- The standardization and coherence among the tools make them easy to learn and use.
- The tools work very well for simple and small problems, but also scale very intuitively and naturally to quite complex and large problems.
- There are tools for every part of the workflow, from data management to statistical analysis and making beautiful graphs.
- Each of us independently migrated to this new set of tools, giving us greater confidence that it's the way forward. (Well, Andrew was forced a bit.)

Though we are confident that teaching newcomers these new tools is the right thing to do, there are some risks and, in particular, people taught only these new tools may not be able to work easily with people or code using

---

[1] Unofficial knighthood for contributions to making our R-lives so much easier and beautiful.

the classic way. Furthermore, some colleagues have questioned the wisdom of teaching this 'modern' approach to entry-level students (i.e. those with no or little previous experience with R), especially if taught in the absence of the classic approach (funnily enough, many of these 'concerned' colleagues don't use R at all!). Certainly the risks mentioned above are real, and for that reason we provide a short appendix in Chapter 3 (the chapter on Data management) that links the classic and new methods. The classic way can still sometimes be the best way. And old dogs don't often agree to learning new tricks.

Another concern voiced asks why we're teaching 'advanced R' at entry level, with the idea that the use of new tools and add-on packages implies 'advanced'. After all, why wouldn't the 'base' R distribution contain everything an entry-level user needs? Well, it does, but we've found the standardization and syntax in the add-on packages to be valuable even for us as seasoned users. And one should not read 'base' R distribution as 'basic' R distribution, or 'add-on' package as 'advanced' package. The 'base' distribution contains many advanced tools, and many add-on packages contain very basic tools.

We hope you enjoy this new *Getting Started with R*.

What this book is about

We love R. We use statistics in our everyday life as researchers and teachers. Sometimes even more: Owen used it to explore the nursing behaviour of his firstborn. We are first and foremost evolutionary and community ecologists, but over the past 15 years we have developed, first in parallel and then together, an affinity for R. We want to share our 40+ years of combined experience using R to show you how easy, important, and exciting it can be. This book is based on 3–5-day courses we give in various guises around the world. The courses are designed to give students and staff alike a boost up the steep initial learning curve associated with R.

We assume that course participants, and you as readers, already use some spreadsheet, statistical, and graphing programs (such as Excel, SPSS,

Minitab, SAS, JMP, Statistica, and SigmaPlot). Most participants, and we hope you, have some grasp of common statistical methods, including the chi-squared test, the *t*-test, and ANOVA. In return for a few days of their lives, we give participants knowledge about how to easily use R, and R only, to manage data, make figures, and do statistics. R changed our research lives, and many participants agree that it has done the same for them.

The efforts we put into developing the course and this book are, how-ever, minuscule compared with the efforts of the R Core Development Team. Please remember to acknowledge them and package contributors when you use R to analyse and publish your amazing findings.

WHAT YOU NEED TO KNOW TO MAKE THIS BOOK WORK FOR YOU

There are a few things that you need to know to make this book, and our ideas, work for you. Many of you already know how to do most of these things, having been in the Internet age for long enough now, but just to be sure:

1. You need to know how to download things from the Internet. If you use Windows, Macintosh, or Linux, the principles are the same, but the details are different. Know your operating system. Know your browser and know your mouse/trackpad.

2. You need to know how to make folders on your computer and save files to them. This is essential for being organized and efficient.

3. It is useful, though not essential, to understand what a 'path' is on your computer. This is the address of a folder or a file (i.e. the path to a file). On Windows, depending on the type you are using, this in-volves a drive name, a colon (:), and slashes (\ or /). On a Macintosh and Linux/Unix, this requires the names of your hard drive, the name of your home directory, a tilde (~), the names of folders, and slashes (/).

4. Finally, you need at least a basic understanding of how to do, and why you are doing, statistics. We recommend that you know the types of questions a *t*-test, a chi-squared test, linear regression, ANOVA, and ANCOVA are designed to help you answer *before* you use this book. As we said, we are not aiming to teach you statistics per se, but how to do some of the most common plotting and most frequent statistics in R, and understand what R is providing as output. That said, we'll try and teach a bit along the way.

How the book is organized

In this book, we will show you how to use R in the context of everyday research in biology (or, indeed, in many other disciplines). Our philosophy assumes that you have some data and would like to derive some understanding from it. Typically you need to manage your data, explore your data (e.g. by plotting it), and then analyse your data. Before any attempt at analysis, we suggest (no, demand!) that you always plot your data. As always, analysing (modelling) your data involves first developing a model that accurately reflects your question, and then testing critical assumptions associated with the statistical method (model). Only after this do you attempt interpretation. Our focus is on developing a rigorous and efficient routine (workflow) and a template for using R for data exploration, visualization, and analysis. We believe that this will give you a functional approach to using R, in which you always have the goal (understanding your data, answering your question) in mind.

Chapter 1 is about getting R and getting acquainted with it. The chapter is a bit like when you first meet someone who might be your friend, or might not, so you take some time to get to know each other. We also introduce you to another friend, RStudio, and strongly recommend that you get to know this one, as well as R. RStudio is just great. You will fall in love with it.

Chapter 2 is about getting your data ready for R, getting it into R, and checking it got into R correctly. Not many courses cover data preparation as much as in this chapter, but it's really essential for an efficient experience with R. Good preparation makes for great performance. We give tips about what can go wrong here, how to recognize this, and how to fix it.

Chapter 3 focuses on how you work with data once it's in R. Usually you'll need to do some data manipulation before making a graph or doing a statistical analysis. You might need to subset your data, or want to calculate mean ± SE. We walk you through some very efficient and clear methods for doing all kinds of data manipulations.

Chapter 4 is about visualizing your data, and comes before the chapters about statistical analyses because *we always visualize our data before we do any statistics* (you will hear that again and again throughout this book). We introduce you to scatterplots, histograms, and box-and-whisker plots. In later chapters, we also introduce you to plots of means and standard errors. (But we do not introduce you to bar charts with error bars, because they are evil[2].)

Chapters 5, 6, and 7 finally get on to doing some statistics. Chapter 5 introduces 'basic' statistical tests (*t*-test, chi-squared contingency table analyses, simple linear regression, and the one-way ANOVA). Chapter 6 is slightly more complex tests (two-way ANOVA and ANCOVA). And Chapter 7 takes us to new territory, where we introduce about the simplest generalized linear model around: a Poisson regression. As we said, we are introducing how to do stuff in R and we're not aiming to cover lots of statistics in great detail, but along the way we try and ensure that your understanding of statistics maps onto the output you can get from using R. We've added this 'getting started with generalized linear models' chapter because so many types of question in the biological sciences demand it. Our goal is that you should have seen enough variety of analysis methods to be comfortable and confident in moving forward and learning more yourself.

Chapter 8 comes back to figures and graphs. It is about how to make your graphs look even more beautiful than they were during the previous

[2] http://dx.doi.org/10.1371/journal.pbio.1002128

chapters. Put another way, it's about pimping your graphs. Making the labels, symbols, colours, shading, sizes, and everything else you might like to change look beautiful, coordinated, and clear, so readers are amazed by the clarity with which they can see your findings. It will also give you the skills and flexibility to make atrocious graphs . . . be careful.

The final chapter 9, wraps all this up and provides encouragement. It is brief. We figure that by this point, you'll have had enough of us, and will be raring to get your own data into R. And that is great, because that is when you'll really solidify your learning.

SOME CONVENTIONS IN THE BOOK

We have attempted to be consistent in the typefaces and colours of text in the book, so that you can easily recognize different types of R command. So the text is rather colourful. Hopefully, the advantages of clarity about what is what will outweigh any concerns you might have about colour choices.

Throughout the book, we highlight where you can work along with us on your own computer using R, through the use of the symbol at the side of the page.

Finally, all of the datasets we use are available online at http://www.r4all.org/the-book/datasets/.

Why R?

If you've got this far, you probably know you want to learn R. Some of you will have established research careers based around using a variety of statistical and graphing packages. Some of you will be starting your research career and wondering whether you should use some of the packages and applications that your supervisor/research group uses, or jump ship to R. Perhaps your group already uses R and you are just looking for that 'getting started' book that answers what you think are embarrassing questions. Regardless of your stage or background, we think an informal but structured introduction to an approach and routine for using R will help. And regardless of the motivation, we finish the Preface here by introducing a core set of features and characteristics of R that we think make it worth using and worth making a transition to from other applications.

First, we think you should invest the effort because it is freely available and cross-platform (e.g. it works on Windows, Macs (OS X), and Linux). This means that no matter where you are and with whom you work, you can share data, figures, analyses, and, most importantly, the instructions (also known as scripts and code) used to generate the figures and analyses. Anyone, anywhere in the world, with any kind of Windows, Macintosh, or Linux operating system, can use R, without a licence. If you, your department, or your university invest heavily in multiple statistical packages, R can save a great deal of money. When you change institutions, R doesn't become inaccessible, get lost, or become unusable.

Second, R is an interpreted programming language. It does not involve extensive use of menus; you type commands instead. As a result, you have to know what to ask R, know why you are asking R for this, and know what to expect from R. You can't just click on menus and get some results. This means that by using R, you continually learn a great deal about statistics and data analysis.

Third, it's free. Oh, we said that already. Actually, it's more accurate to state that it's freely available. Lots of people put an awful lot of effort into developing R . . . that effort wasn't free. Please acknowledge this effort by citing R when you use it.

Fourth, we believe that R can replace common combinations of programs that you might use in the process of analysing your data. For example, we have, at times, used two or three of Excel, Minitab, SAS, Systat, JMP, SigmaPlot, and CricketGraph, to name a few. This results in not only costly licensing of multiple programs, but also software-specific files of various formats, all floating around in various places on your computer (or desk) that are necessary for the exploration, plotting, and analysis that make up a research project. Keeping a research project organized is hard enough without having to manage multiple files and file types, proprietary data formats, and the tools to put them all together. Furthermore, moving data between applications introduces extra steps into your workflow. And how much fun is it piecing all of this together 3–6 months after submitting a manuscript, and needing to make changes? These steps and frustrations are removed by investing in using R.

Fifth, with R you can make outstanding publication-quality and publication-ready figures, and export them in many different formats, including pdf. We now use only R for making graphs, and when submitting manuscripts to journals we usually send only pdf files generated directly from R. One of the nice things about pdfs is that they are resolution independent (you can zoom in as far as you like and they don't get blocky). This means that publishers have the best possible version of your figure. And if the quality is poor in the published version of your paper, you know it is down to something the publishers have done!

Finally, and quite importantly, R makes it very easy to write down and save the instructions you want R to execute—this is called a script in R. In fact, the script becomes a permanent, repeatable, annotated, cross-platform, shareable record of your analysis. Your entire analysis, from transferring your data from field or lab notebook to making figures and performing analyses, is all in one, secure, repeatable, annotated place.

Take your time and learn the magic of R. Let's get started.

Updates

Rstudio evolves quickly, so don't worry if what you see on your computer is a little different from what's printed in this book. For example, as this book went to press, RStudio started using a new method for importing data. We quickly updated the most important parts of the book, but for a full account of this change, and any others, look on the book web site `www.r4all.org/the-book`.

Acknowledgements

Thanks to our wives, Sophie, Amanda, and Sara, for everything. After all these years, they know about R too. Many thanks to Ian Sherman and Lucy Nash at OUP for their guidance, support and encouragement, to Douglas Meekison for excellent copy-editing, and Philip Alexander for patiently dealing with countless "final" fixes!

1

Getting and Getting Acquainted with R

1.1 Getting started

One of the most challenging bits of getting started with R is actually getting R, installing it, and understanding how it works with your computer. Despite R's cross-platform capacity (OSX, Windows, Linux, Unix), there remain several differences in how things can look on each platform. Thankfully, a new application, RStudio, provides a way to standardize most of what you see and do with R, once it is on your computer. In this chapter, we'll walk you through the steps of getting R and RStudio, installing them on your computer, understanding what you've done, and then working through various aspects of using R and RStudio.

This introduction will make you feel comfortable using R, via RStudio. It will make you understand that R is a giant calculator that does whatever you ask it to do (within reason). It will also familiarize you with how R does things, both 'out of the box' and via additional 'add-on' packages that make R one of the most fun and widely used programs for doing statistics and visualizing data.

Getting Started with R Second Edition. Andrew Beckerman, Dylan Childs, & Owen Petchey: Oxford University Press (2017). © Andrew Beckerman, Dylan Childs, & Owen Petchey. DOI 10.1093/oso/9780198787839.001.0001

We will first walk you through getting and installing R and getting and installing RStudio. While for many this will be trivial, our experience suggests that many of you probably need a tiny bit of hand-holding every once and a while.

1.2 Getting R

We assume you don't yet have R on your computer. It will run on Macintosh, Windows, Linux, and Unix operating systems. R has a homepage, `r-project.org`, but the software itself is located for download on the Comprehensive R Archive Network (CRAN), which you can find at `cran.r-project.org` (Figure 1.1).

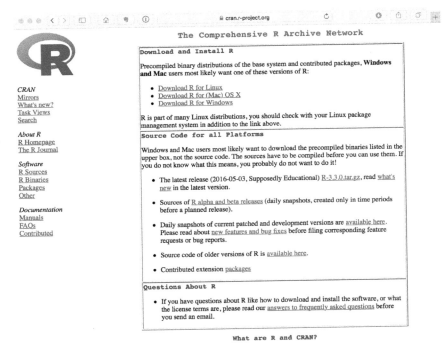

Figure 1.1   The CRAN website front page, from where you can find the links to download the R application.

The top box on CRAN provides access to the three major classes of operating systems. Simply click on the link for your operating system. As we mentioned in the Preface, R remains freely available.

You'll hear our next recommendation quite a bit throughout the book: *read the instructions*. The instructions will take you through the processes of downloading R and installing it on your computer. It might also make sense to examine some of the Frequently Asked Questions found at the bottom of the web page. R has been around quite a long time now, and these FAQs reflect more than a decade of beginners like you asking questions about how R works, etc. Go on . . . have a look!

1.2.1 LINUX/UNIX

Moving along now, the Linux link takes you to several folders for flavours of Linux and Unix. Within each of those is a set of instructions. We'll assume that if you know enough to have a Linux or Unix machine under your fine fingertips, you can follow these instructions and take advantage of the various tools.

1.2.2 WINDOWS

The Windows link takes you to a page with three more links. The link you want to focus on is 'base'. You will also notice that there is a link to the aforementioned R FAQs and an additional R for Windows FAQs. Go on . . . have a look! There is a tonne of good stuff in there about the various ways R works on Windows NT, Vista, 8, 10, etc. The base link moves you further on to instructions and the installer, as shown in Figure 1.2.

1.2.3 MACINTOSH

The (Mac) OS X link takes you to a page with several links as well (Figure 1.3). Unless you are on a super-old machine, the first link is the one on which you want to focus. It will download the latest version of R for several recent distributions of OS X and offer, via a .dmg installer, to put everything where it needs to be. Note that while not required for 'getting started', getting the XQuartz X11 windowing system is a good idea;

Windows Step 1   Windows Step 2

Figure 1.2 Two steps to download the Windows version of R.

R for Mac OS X

This directory contains binaries for a base distribution and packages to run on Mac OS X (release 10.6 and above). Mac OS 8.6 to 9.2 (and Mac OS X 10.1) are no longer supported but you can find the last supported release of R for these systems (which is R 1.7.1) here. Releases for old Mac OS X systems (through Mac OS X 10.5) and PowerPC Macs can be found in the old directory.

CRAN
Mirrors
What's new?
Task Views
Search

About R
R Homepage
The R Journal

Software
R Sources
R Binaries
Packages
Other

Documentation
Manuals
FAQs
Contributed

Note: CRAN does not have Mac OS X systems and cannot check these binaries for viruses. Although we take precautions when assembling binaries, please use the normal precautions with downloaded executables.

As of 2016/03/01 package binaries for R versions older than 2.12.0 are only available from the CRAN archive so users of such versions should adjust the CRAN mirror setting accordingly.

R 3.3.0 "Supposedly Educational" released on 2016/05/03

Please check the MD5 checksum of the downloaded image to ensure that it has not been tampered with or corrupted during the mirroring process. For example type
md5 R-3.3.0.pkg
in the *Terminal* application to print the MD5 checksum for the R-3.3.0.pkg image. On Mac OS X 10.7 and later you can also validate the signature using
pkgutil --check-signature R-3.3.0.pkg

Files:

R-3.3.0.pkg
MD5-hash: 871f274d0f9a12731d328125774c639a
SHA1-
hash: a0ac68233cb826d228d7791e994fc634a888cc48
(ca. 71MB)

R 3.3.0 binary for Mac OS X 10.9 (Mavericks) and higher, signed package. Contains R 3.3.0 framework, R.app GUI 1.68 in 64-bit for Intel Macs, Tcl/Tk 8.6.0 X11 libraries and Texinfo 5.2. The latter two components are optional and can be ommitted when choosing "custom install", it is only needed if you want to use the tcltk R package or build package documentation from sources.

Note: the use of X11 (including tcltk) requires XQuartz to be installed since it is no

Figure 1.3 The download page for R for Macintosh.

a link is provided just below the paragraph describing the installer (see Figure 1.3). As with Windows, the R FAQs and an additional R for OS X FAQs are provided . . . they are *good things*.

1.3 Getting RStudio

So, at this stage, you should have downloaded and installed R. Well done! However, we are not going to use R directly. Our experience suggests that you will enjoy your R-life a lot more if you interact with R via a different program, also freely available: the software application *RStudio*. RStudio is a lovely, cross-platform application that makes interacting with R quite a bit easier and more pleasurable. Among other things, it makes importing data a breeze, has a standardized look and feel on all platforms, and has several tools that make it much easier to keep track of the instructions you have to give R to make the magic happen.

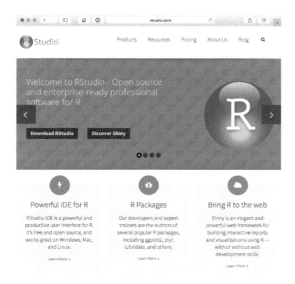

Figure 1.4 The RStudio website front page, from where you can find the links to download the RStudio application. (Note: you must (as you have done) also download the R application from the CRAN website.)

We highly recommend you use RStudio to get started (we use it in teaching and in our research; Figure 1.4). You can read all about it here: `https://www.rstudio.com`. You can download RStudio here: `https://www.rstudio.com/products/rstudio/download/`.

At this point, you should have downloaded and installed R, and downloaded and installed RStudio.

1.4 Let's play

You are now ready to start interacting with R. RStudio is the application we will use. In the process of installing RStudio, it went around your hard drive to find the R installation. It *knows* where R is. All we need to do now is fire up RStudio.

Start RStudio. Of course, *you* need to know where it is, but we assume you know how to find applications via a 'Start' menu, or in the Applications folder or via an icon on the desktop or in a dock . . . however you do this, navigate to RStudio, and start it up. You are clever. You know how to start an application on your computer!

When we teach, several people end up opening R rather than RStudio. The RStudio icon looks different from the R icon. Make sure you are starting RStudio (Figure 1.5). If all has gone to plan, the RStudio application

Figure 1.5 The R and RStudio icons are different. You want to be using the RStudio application.

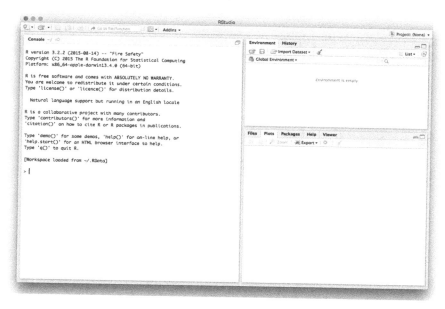

Figure 1.6 The RStudio application initiates the first time you open it with three *panes*. Left is the *Console*; top right is a pane with two tabs, *Environment* and *History*; and bottom right is a pane with five tabs, *Files*, *Plots*, *Packages*, *Help*, and *Viewer*. These are explained in detail in the main text.

will, at its very first start, give you three *panes* (Figure 1.6).[1] Let's walk through them.

On the left is the *Console* pane. This is the window that looks into the engine of R. This is the window where you can give instructions to R, they are worked on by the little people inside R that are really smart and talk in a language of 1s and 0s, and then the answer appears back in the Console pane. It is the brain. The mind. The engine.

The top right is a pane with two tabs: *Environment* and *History*. The Environment pane shows the things that R has in its head. This could be

---

[1] If you have used RStudio already, it might be showing four panes; don't worry. In what follows, work with the lower left pane, the *console*.

datasets, models, etc. It will probably be empty at the moment, but you will soon start to fill it up. The Environment pane also contains a very important button: *Import Dataset*. We will cover the use of this extensively in Chapter 3. The History pane contains the instructions R has run.

Bottom right is a pane with five tabs: *Files*, *Plots*, *Packages*, *Help*, and *Viewer*. They are rather self-explanatory, and as we begin to use RStudio as our interface for R, feel free to watch what shows up in each.

When you start RStudio, the Console gives some useful information about R, its open source status, etc. But, for new users, the most important is at the bottom, where you will see the symbol > with a cursor flashing after it. This is known as the *prompt*.

You can only type instructions in one place in the Console, and that is at the prompt. Try giving R your first instruction. Click in the Console and type 1 + 1 and then press enter/return. You should see something like this (though there will not be two # characters at the start of the answer line):

```
1 + 1

## [1] 2
```

The instruction we gave to R was a question: 'Please, can you give us the answer to what is one plus one?' and R has given back the answer

```
[1]  2
```

You can read this as R telling you that the first (and only, in this case) part of the answer is 2. The fact that it's the first part (indicated by the one in square brackets) is redundant here, since your question only has a one-part answer. Some answers have more than one part.

After the answer is a new line with the prompt. R is again ready for an instruction/question.

1.5 Using R as a *giant* calculator (the size of your computer)

What else can R do, other than add one and one? It is a giant calculator, the size of your computer. As befits a statistical programming language, it can divide, multiply, add, and subtract; it conforms to this basic order too (DMAS). It can also raise to powers, log numbers, do trigonometry, solve systems of differential equations … and lots of other maths. Here are some simple examples. Let's go ahead and type each of these into the Console, pressing enter/return after each to see the answer:

```
2 * 4
## [1] 8
3/8
## [1] 0.375
11.75 - 4.813
## [1] 6.937
10^2
## [1] 100
log(10)
## [1] 2.302585
log10(10)
## [1] 1
sin(2 * pi)
## [1] -2.449294e-16
7 < 10
## [1] TRUE
```

* In these blocks of R in this book the ## lines are answers from R. Don't type them in.

Pretty nice. There are a few things worth noting here—some 'default' behaviours hard-wired into R. These are important, because not all statistical software or spreadsheet software like Excel handles things the same way:

- If you understand and use logarithms, you might be puzzled by the result of log(10), which gives 2.30. In R, **log**(x) gives the natural log of x, and not the log to base 10. This is different from other software, which often uses `ln()` to give the natural log. In R, to make a log to base 10, use **log10**(x). See, in the example, **log10**(10) = 1. You can use **log2**() for the log to base 2.
- The trigonometric function **sin**() works in radians (not degrees) in R. So a full circle is $2 \times \pi$ radians (not 360 degrees).
- Some mathematical constants, such as π, are built into R.
- The answer to **sin**(2*pi) should be zero, but R tells us it is very close to zero but not zero. This is computer mumbo-jumbo. The people that

built R understand how computers work, so they made a function called **sinpi**() that does the 'multiply by π' bit for you—**sinpi**(2) does equal zero.

- We sometimes didn't include any spaces in the instructions (e.g. there were no spaces around pi). It would not have mattered if we had, however. R ignores such white space. It ignores all white space (spaces, new lines, and tabs).
- The last question is 'is 7 less than 10?' R gets this right, with the answer 'TRUE'. The 'less than' sign is known as a 'logical operator'. Others include == (are two things equal?), ! = (are two things not equal?), > (is the thing on the left greater than the thing on the right?), <= (less than or equal to), >= (greater than or equal to), | (the vertical bar symbol, not a big i or a little L; is one or the other thing true?), and & (are two things both true?).

If you were watching carefully too, you will have noticed that RStudio is very nice to you, pre-placing closing brackets/parentheses where necessary. Super-nice.

We've also just introduced you to a new concept: functions like **log10**(), **log**(), and **sin**(). Box 1.1 explains more about what functions are. Dip into it at your pleasure!

1.5.1 FROM THE SIMPLE TO THE SLIGHTLY MORE COMPLEX

All of the maths above gave just one answer. It's the kind of maths you're probably used to. But R can answer several questions at once. For example, we can ask R 'Would you please give us the integers from 1 to 10, inclusive?' We can do this two ways. Let's start with the easy way:

```
1:10

##   [1]   1   2   3   4   5   6   7   8   9  10
```

The answer to our question has ten elements. But hang on, there is only [1] in front of everything. Fret not, as R is being brief here. It has limited the square brackets to the left-hand side of the screen and not told us that 2 is the second answer ([2]), 3 is the third ([3]), and so on. R assumes you're clever! You can even try extending the sequence to 50, and see how R inserts the addresses [ ] only on the left-hand side of the Console.

Box 1.1: But what are functions?

Hopefully you're starting to feel that R isn't so difficult after all. But you'll almost certainly have questions. One might be *what are functions?* A function introduced on the next page is **seq**() (make sure you can find this in the text below). Asking R to do things usually requires using functions. R uses functions to do all kinds of things and return information to you. All functions in R are a word, or a combination of words containing no spaces, followed by an opening bracket '(' and a closing bracket ')'. Inside these brackets goes the information that we give to the function. These bits of information are called 'arguments'—yes, sometimes it feels like arguing.

Arguments are separated by commas. We introduce the **seq**() function below to make a series of numbers:

```
seq(from = 0, to = 10, by = 1)
```

The function is **seq**(), and inside the function brackets are three arguments separated by two commas (necessarily). The first argument is the value to use at the start of the sequence, the second is the value to use at the end of the sequence, and the third is the step size.

To clear R's brain, we use a function inside another function: **ls**() inside **rm**(). The **rm**() stands for remove, and **ls**() stands for list. We combine them in the following way to make R 'clear' its brain:

```
rm(list = ls())
```

It is best to read this from the inside out. The **ls**() requests all of the objects in R's brain. The **rm**() asks R to remove all of these objects in the list. The `list=` is new, and is us telling R exactly what information (i.e. which argument) we are giving to the **rm**() function. You will have noticed that you now know a function that gives a list of all of the objects in R's brain: **ls**(). This is a handy function.

We will introduce and explain many more functions, as they're the workhorses of R. We'll also repeat some of this information about functions (because it's so important), and also explain some things in more detail (e.g. why sometimes we explicitly tell a function the information we are giving it, and why sometimes we can get away without doing so).

The : in 1:10 tells R to make a sequence of whole numbers that goes up in steps of one. We can also generate this sequence using a function that R has in its toolbox. It is called ... wait for it ... seq(). Wow! *Rocket Science!*

1.5.2 FUNCTIONS TAKE ARGUMENTS

seq() is a function, and in R, functions do clever things for us to make life easier. But we have to give functions things called arguments to control what they do. This isn't complicated. Let's look closely at how we use seq(). We have to provide three arguments to seq(): the first value of the sequence, the last value of the sequence, and the step size between numbers (the difference in value between numbers in the sequence). For example:

```
seq(from = 1, to = 10, by = 1)

##  [1]  1  2  3  4  5  6  7  8  9 10
```

This reads 'Please give us the sequence of numbers that begins at 1, ends at 10, and has a 1 unit difference between the numbers.' Formally, the arguments are called *from*, *to*, and *by*. We suggest you *do* write the names of these arguments as you use functions. They are not required, but without naming your arguments, you risk getting strange answers if you put things in the wrong place, or, worse, a dreaded *red* error message.

And the answer is what we would expect. Note that we have included some spaces in the instruction; specifically, we have put one space at each comma. R doesn't care; it would be just as happy with no spaces. We used the spaces so the instruction is easier for you to read. More generally, you should attempt to write instructions that will be easier for the two most important readers of your instructions: you and other people. It's easy to focus on writing instructions that R can read. We really should also focus on writing instructions that are easy for humans to read, and for us to read in six months' time when we have to revise our amazing manuscript.

Let's now modify the our use of **seq()** to provide a sequence from 1 to 10 in steps of 0.5:

```
seq(from = 1, to = 10, by = 0.5)
```

```
##  [1]  1.0  1.5  2.0  2.5  3.0  3.5  4.0  4.5  5.0  5.5  6.0
## [12]  6.5  7.0  7.5  8.0  8.5  9.0  9.5 10.0
```

The answers are those we'd expect. Note that we see above that we now get an address ([12]) on the second line of answers. This is R being helpful ... it's giving us a clue about which answer we've got to by the second line, i.e. 6.5 is the 12th answer. Here's what happens if we make the R Console narrower:

```
seq(from = 1, to = 10, by = 0.5)
```

```
##  [1]  1.0  1.5  2.0  2.5  3.0  3.5  4.0
##  [8]  4.5  5.0  5.5  6.0  6.5  7.0  7.5
## [15]  8.0  8.5  9.0  9.5 10.0
```

The question and answer are the same. But we have made the answer go over three lines, and R tells us on each new line the number or address of the answer it's reporting at the beginning of each line.

1.5.3 NOW FOR SOMETHING REALLY IMPORTANT

So far, R has printed the answer to questions in the Console. R hasn't kept the answer in its head or saved it anywhere. As a result, we can't do anything else with the answer. Effectively, it is gone from R's head. It might even be gone from your head!

Often we will want to use the answer to one question in a subsequent question. In this case, it is convenient, if not essential, to have R keep the answer. To have R do this, we *assign* the answer of a question to an *object*. Like this, for example:

```
x <- seq(from = 1, to = 10, by = 0.5)
```

A few things to note:

- We do the assignment by using the *assignment arrow*, which is a less than sign followed (without a space) by a minus sign: < -. The arrow points from right to left, so the assignment goes from right to left.

- We assign the answer to something called x. We could have used more or less anything (any continuous text string that starts with a letter, more precisely) that we wanted here. We encourage you to be sensible: we will probably have to type this object name again, so don't make it too long. (But) make it informative if possible.
- After we press enter/return, a new prompt appears in the next line. We don't see the answer to our question, because R has assigned the values produced on the right-hand side to the thing on the left-hand side. That is, if R is able to interpret our instruction, it does what it is told and then just says it's ready for the next instruction. R doesn't congratulate us when we are successful; it only criticizes us when we make a mistake. Get used to this!

To see the answer to the question is easy: just type x in the Console and press return:

```
x
```

```
##  [1]  1.0  1.5  2.0  2.5  3.0  3.5  4.0  4.5  5.0  5.5  6.0
## [12]  6.5  7.0  7.5  8.0  8.5  9.0  9.5 10.0
```

1.5.4 HOW DOES R DO STUFF WITH VECTORS?

Now, try asking R to give you the numbers from 101 to 110 in steps of 0.5 and assign these to an object called y. Then, add together the objects x and y. It *is* as easy as you think ... but before you do that, think about what you expect the answer will be.

Here's what you could/should have done:

```
y <- seq(from = 101, to = 110, by = 0.5)
x + y
```

```
##  [1] 102 103 104 105 106 107 108 109 110 111 112 113
```

```
## [13] 114 115 116 117 118 119 120
```

Great, you just added together two vectors (these two vectors are both a collection of numbers) and thus experienced a hint of the power of R! But wait ... there is a bit more. We think it is pretty cool to note just how R did the maths here. If you look and think carefully, you will realize that R added the vectors element by element. This is very cool. All those sums, all at once.

Now, before you discover more of this awesome power, we think it's time you learn to *never again* type instructions into the Console. Well, almost never.

1.6 Your first script

Up to know, you have been typing instructions into the Console, pressing enter, and watching R give you the answer (or assigned the answer to an object if this is what you asked). We *very* strongly advise you not to regularly do this. In fact, you will have real trouble if you do this regularly. Perhaps the only consistent use of typing in the Console should be asking for help (see below).

The alternative is to type your instructions in a separate place, and then send them to the Console (on a magic carpet) when you want R to work on them. As we noted in the Preface, R allows you to write down and save the instructions that R then uses to complete your analysis. Trust us, this is a desirable feature of R. As a result of making a *script*, you, as a researcher, end up with a *permanent, repeatable, annotated, shareable, cross-platform* archive of your analysis. Your entire analysis, from your raw, transferred data from lab book to making figures and performing analyses, is all in one, *secure, repeatable, annotated* place. We think you can see the value. We continue to benefit from this every time we get referees' comments back from a journal . . .

Because this is such a good way of working with R, both the basic version of R for Windows and Mac, and RStudio for all platforms have built-in *text editors*. These are a bit like a word-processing pane that you write stuff in, but these editors tend to have at least one major feature that word processors don't—there is a keystroke combination that, when pressed

together, magically sends information from the script to the Console
window. This is very convenient; much more so than copying and pasting.

1.6.1 THE SCRIPT PANE

Let's first find the script pane in RStudio, and then explore this and two
other features we think make it amazing. To see the script pane, you need
to look in the upper right corner of the Console pane that is dominating
your screen. Up there, on the right, you should see two squares overlapping
each other. Go on. Press them. All of a sudden, like magic (we all need
magic), there should be two panes on the left. The top pane is your *script*
pane (Figure 1.7). From now on, when you open RStudio, there will be
four panes, not three.

The script editor in RStudio has some nice features *not* found in several
other versions of the R application. To see some of these in action, let's get
started with a new script. To open a new script, go to the *File* menu, then
New File, and click on *R Script*. Or you can click on the top left button

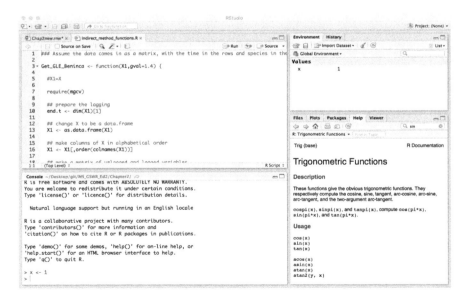

Figure 1.7   How the RStudio application looks with four panes . . . the new one
is the script pane.

in the script toolbar, which looks like a document with a green and white plus, and then click on *R Script* in the drop-down menu.

You now have somewhere to write your instructions (and save them). Let's start writing in the script with a very special symbol, the # symbol. # is a very special character. It is important to find this character on your keyboard. It may not be obvious on European keyboards, especially European Macintosh keyboards. If it is not visible, it is often associated, somehow, with the number 3 (for example, alt+3 on many Macintosh keyboards). The 3 often has the £ symbol on it as well. The # symbol is sometimes called the 'pound sign', like in parts of the USA.

The # is a symbol that declares that anything after it on that line is a comment, i.e. an annotation. These are words for you, not for R. We recommend starting your scripts with some information about the script, the date, and the author ... For example,

```
# Amazing R. User (your name) 12 January, 2021 This script is
# for the analysis of coffee consumption and burger eating
```

One thing you might notice is that this text is shown in a different colour from black (probably in green). This is good. Hold onto whatever you are thinking ... Next in our script, we recommend adding two lines as follows:

```
# clear R's brain
rm(list = ls())
```

This bit of R magic is very important, and you should have it at the beginning of virtually every script you write. It clears R's brain. It's always good to start a script with nothing accidentally left over in R from a previous session. Why? Well, imagine that you're analysing the body size of your special species in two different experiments. And, not surprisingly, you've called it **body_size** in both datasets. You finish one analysis, and then get started on the next. By clearing R's brain before you start the second analysis, you are ensuring that you don't use the wrong body size data for an analysis.

After this bit of R code, you can now add a bit more annotation/commentary, and some interesting maths operations for R to do, like this:

```
# Amazing R. User (your name)
# 12 January, 2021
# This script is for the analysis of coffee consumption and
# burger eating

# Clear R's brain
rm(list = ls())

# Some interesting maths in R
1 + 1
2 * 4
3 / 8
11.75 - 4.813
10^2
log(10) # remember that log is natural in R!
log10(10)
sin(2*pi)
x <- seq(1, 10, 0.5)
y <- seq(101, 110, 0.5)
x + y
```

Here we highlight some things to note about this script:

- The first four lines begin with a hash (sometimes called the pound sign), '#'. This is the symbol that tells R to ignore everything that follows on the line. So R doesn't even read the 'Amazing R. User (your name)' text, which is good, because it wouldn't have the first idea what this means. The # means what follows is for humans, not for R.
- The script contains only the instructions for R. It does not contain the answers. We still have to get R to calculate these for us.
- The seventh line (including the empty one) is very important, and you should have it at the beginning of virtually every script you write. Did we mention that already?

- Note that there are at least four colours being used in your script. This is called syntax highlighting. This is a good thing. It separates comments, R function, numbers, and other things.
- If you were watching carefully as you typed, you will have noticed that RStudio was completing your brackets/parentheses for you . . . every time you opened one up, RStudio provided the closing one. *Very handy.*
- It has white space. Between the first annotation and the Brain Clearing (line 7), there was a line with nothing. After the brain clearing, there was a blank line. We recommend this kind of white space. It creates *chunks* of code and annotation. It makes your script easy to read, and easy to follow; it is good practice.

Now, have a look up at the colour of the word '*untitled1*' in the tab of the script pane. It is probably *red*. That's bad. Warning colour. *Danger.* Your work is not saved. Now it is time to save this script to your analysis folder. You can either choose File -> Save or use keystrokes like ctrl+S (Windows) or cmd+S (Macintosh). Provide an informative name for the script. Save it. Don't lose all the hard work.

OK. Danger has passed. Breathe easy. The next bit is fun.

1.6.2 HOW DO I MAKE R DO STUFF, YOU ASK?

RStudio makes it easy to get these (or any) instructions from the script editor (source) to the Console. The hard, slow, and boring way to do this is to select/highlight with your mouse or keyboard or trackpad the text you want to put in the Console, copy it, then click in the Console, paste it, and then press enter to run the script. Phew . . .

But there is magic in these computer things. The super-easy, quick, and exciting method is to click anywhere in the line you want to submit (you don't have to highlight the whole line) and press the correct keyboard shortcut. On a Mac press cmd+enter OR ctrl+enter, and on Windows ctrl+enter. You will then see the line of code magically appear in the Console. Nice!

If you insist on using the mouse, in RStudio you can also press the Run button, found on the upper right of the script tab (it actually says 'Run') to send script lines from the editor to the Console. We think it is faster, however, to use keyboard shortcuts. Most of these shortcuts are actually documented in the Code and View menu items in RStudio (Figure 1.8) and on the RStudio website (google for RStudio keyboard shortcuts; probably the first link is highly relevant).

Furthermore, if you highlight several lines of code in the script editor and press the shortcut, all of them are delivered/submitted to the Console in one go. And to highlight and submit all of the instructions in a file of script, press cmd+a (Mac) or ctrl+a (Windows) to select everything in the script, then use the shortcut to submit those instructions.

Please, please, please use these shortcuts for getting instructions from the script editor to the Console. If you ever find yourself copy and pasting from the script editor to the Console, feel bad. Very bad. You have forsaken the magic.

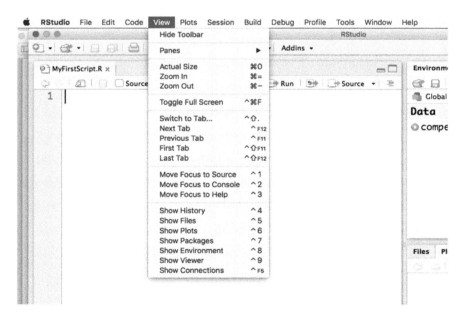

Figure 1.8 The RStudio View menu shows you many, many shortcuts via the keyboard. Check out the Code menu for more.

1.6.3 TWO MORE BITS OF RSTUDIO MAGIC

Let's see two more bits of handy Rstudio magic. First, click in the Console, so that the cursor is flashing at the prompt. Now, press crtl+1. Can you find the cursor? Has it moved up to the Source pane and your script? Now, try crtl+2. Has it moved back? Brilliant. Another quick way to navigate between the two, if needed. Check out the View menu (Figure 1.8) for how to use crtl+1:9!

Second, let's look at something in the Tools menu. Choose Tools -> Global Options. In the box that comes up, select Appearance. Oooh, this looks fun. Not only can you change the font, but you can also alter the way things are coloured. Recall we noted above that the script had at least four colours. Go ahead and play around with the 'editor theme'. One of us is a particular fan of Tomorrow Night 80's. Go figure. Don't forget to press 'Apply' and then 'OK' to make these changes. Just one way to take control . . .

1.7 Intermezzo remarks

At this point, you have the tools to start engaging with R. You should mess around with RStudio and make it your own. Don't let it control you . . . be in charge. You are also aware now of how powerful the use of a script can be. We want to emphasize that it offers something extraordinary. As a result of making a script, you, as a researcher, end up with a *permanent, repeatable, annotated, shareable, cross-platform* archive of your analysis. Your entire analysis, from your raw, transferred data from lab book to making figures and performing analyses, is all organized in one, *secure, repeatable, annotated* place.

1.8 Important functionality: packages

R's functionality is distributed among many *packages*. Each has a certain focus; for example, the **stats** package contains functions that apply common statistical methods, and the **graphics** package has functions concerning plotting. When you download R, you automatically get a set of *base* packages. These are *mature* packages that contain widely used statistical and plotting functionality. These base R packages represent a small subset of all the packages you can use with R. In fact, at the time of writing, there are more than 8000. These other packages we call *add-on packages*, because you have to add them to R, from CRAN, yourself.

R packages can be installed in a number of different ways. But, as we might expect, RStudio gives you a nice way of doing this. The *Packages* tab in the bottom right pane has an Install button at the top left. Clicking on this brings up a small window with three main fields: *Install from*, *Packages*, and *Install to Library*. You only need to really worry about the *Packages* field; the other two can almost always be left at their defaults.

When you start typing in the first few letters of a package name (e.g. *dplyr*), RStudio will provide a list of available packages that match this. It is totally possible to install more than one package, placing a comma or a space between the different ones you ask for. After we find them, all we need to do is click the *Install* button and let RStudio do its magic.

Let's do this now. This book is going to use a great deal of two add-on packages: *dplyr* and **ggplot2**. We'd like you to get them and install them on your computer. Follow the instructions above. It should be painless. Once you're done, take a quick look at the Console. If it all worked, you will see that all RStudio did was send the R function **install.packages**() to the Console to install the packages for you.

1.8.1 USING THE NEW FUNCTIONS THAT COME IN A PACKAGE

Once we've installed a package onto our computer, we still have to load it into R's brain. A good way of thinking about all of this is by analogy with your phone (we assume you have some kind of smartphone; apologies if you don't). When you download an app, it's grabbed from the app store and placed on your phone. What you did above was to download to your computer and install in R the *apps* ***dplyr*** and ***ggplot2***. The R app store is CRAN.

However, just like with your phone, these apps don't *just start*. You need to put a finger on the icon to make it start working on your phone. In R, the way we *press the icon* is to use a function called **library()**. Now that you've installed ***dplyr*** and ***ggplot2***, let's add info to your script so that when you run these lines, the packages are *activated* and ready to use!

```
# Amazing R. User (your name)
# 12 January, 2021
# This script is for the analysis of coffee consumption and
# burger eating

# make these packages and their associated functions
# available to use in this script
library(dplyr)
library(ggplot2)

# clear R's brain
rm(list = ls())

# Some interesting maths in R
1+1
2*4
3/8
11.75 - 4.813
10^2
log(10)
log10(10)
sin(2*pi)
x <- seq(1, 10, 0.5)
y <- seq(101, 110, 0.5)
x+y
```

- *Top Tip 1.* Put the **library()** commands at the top of your script, all in one place. This will help you see what you, or someone else, need to have installed in order for the script to run successfully.
- *Top Tip 2.* Just like on your phone, you do not need to *install* the packages every time you start a new R session. Once you have a copy of a package on your hard drive it will remain there for you to use, unless you delete it or you install a new version of R.

- *Top Tip 3.* Note that **rm(list = ls())** does not remove packages (we have placed this code after the two **library()** commands). The brain clearing only removes objects you made. That's why it can come after the **library()** commands.

1.9 Getting help

You are nearly ready to be let loose with R and RStudio. But what about getting help from R?

The classic way is to type in the Console '?the-function-name'. This will open a window containing R's help information about the function. For example, ?**read.csv()** will give you the help file for the **read.csv()** function. Box 1.2 provides insight into how to read the **seq()** help file. Not easy, eh! These help files take no prisoners, but after some practice are very useful. Take the time to go through them.

There are lots of other ways of getting help. Google is a great friend, as usual. Search on Google for something and add the letter R, and you will likely get some very useful results. For example, try searching for 'how to make a scatterplot R'. Probably you will have trouble deciding which of the probably very useful first few results to use.

For more controllable searching of R resources, consider using the R channel on Stack Overflow (`http://stackoverflow.com/tags/r/info`) or RSeek (`rseek.org`), whose search results are from Google, but can be easily filtered by categories such as Support, Books, Articles, Packages, and For Beginners.

Another very nice resource is cheat sheets. These are concise and dense compilations of many common and useful functions/tasks. Excellent examples are the RStudio cheat sheets, which you can access from `https://www.RStudio.com/resources/cheatsheets/` or via the Help menu in RStudio. You will find the *Data Wrangling* and *Data Visualization* cheat sheets particularly useful. Not to mention the one for RStudio itself. We've printed these, laminated them, and have them close to hand at all times. Our families think this a bit weird; we don't care. We know we are a bit weird.

Box 1.2: Getting help from an R help file

There are lots of ways to get help, from asking the person sitting next to you, to consulting Google and to reading books. However, the R help files are very important and very useful for R. They have a consistent formal structure: they always start with the name of the function and the package to which it belongs and a short description. They always end with some examples. In the middle, they always contain sections called *Usage, Arguments, Details, Value, Authors, References, See Also*, and *Examples*.

We can see that **seq()** (Figure 1.9) is a function in the base package of R (upper left). We can use it for 'Sequence Generation'. The description is followed by Usage and Arguments. Note that there is specific usage for a data.frame. There are five main arguments for the method. For example, the *from* argument is the number we'd like

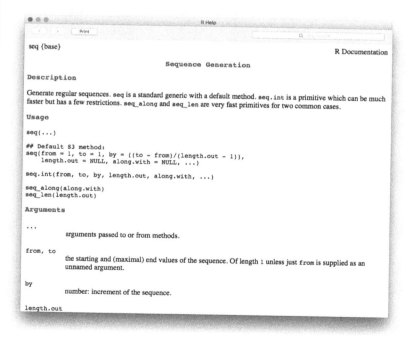

Figure 1.9 The seq help file you can access with ?seq.

Box 1.2: (continued)

R to start the sequence at. We have used from, to, and by. We draw your attention to length.out, which can be shortened to *length*. It is a valuable alternative to by, returning a fixed number of values between the start and finish numbers.

If you scroll down to the bottom of the help file, you will see quite a lot of information in the *Details* section. Then there is information on the *Value* returned by **seq**(), some information on the *Authors*, and information on the *References* that influenced the development of the function. Then comes a section *See Also*, in which related functions are listed (quite useful if the function you're looking at seems not to do exactly what you'd like, and you suspect there might be a more useful function). Finally, and importantly, there are also *Examples*. Note the often prolific use of annotation in the examples. Note as well that you can copy any of the examples and paste them into your Console and they will run. This allows you to dissect the use and anatomy of this function.

The R help files are not always so easy to understand, but often contain answers. So stick with them, with some determination to figure them out; the effort will be well spent.

1.10 A mini-practical—some in-depth play

OK. Let's see what you can do. To get a little more comfortable with R, using the script in RStudio, reading and using help files, and to challenge you, we offer you the chance to try a few exercises (the solutions are in Appendix 1a at the end of this chapter):

- Plot a graph with x^2 on the y-axis and x on the x-axis.
- Plot a graph with the sine of x on the y-axis and x on the x-axis.
- Plot a histogram of 1000 random normal deviates.

The answers are in Figures 1.10–1.12, so you can see what we're aiming for. We'll be using two new functions to solve these exercises: **qplot**() and **rnorm**(). **qplot**() makes a plot, and **rnorm**() gives us random numbers from a normal distribution! Don't forget the '?' to look at help files!

Please, please, please take at least half an hour to attempt these exercises. This is really important. It doesn't matter if you can't get exactly the graphs in Figures 1.10–1.12, but it's really important that you try. In trying, you will experiment with R, get frustrated, get elated (hopefully), experience errors, and have questions. Experiencing all this now will pay dividends later. So it's in your very best interests to stop reading now and have a go. We provide a version of the solutions in Appendix 1a.

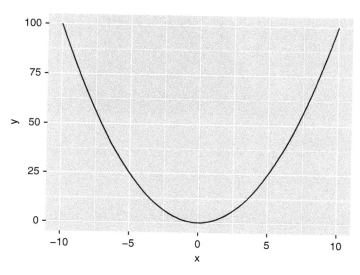

Figure 1.10   The solution to the first problem.

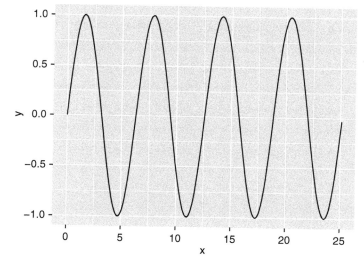

Figure 1.11   The solution to the second problem.

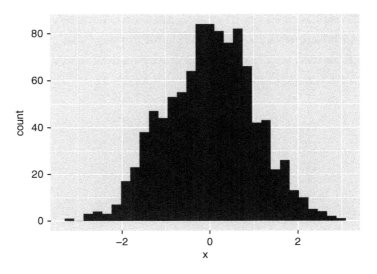

Figure 1.12 The solution to the third problem.

1.11 Some more top tips and hints for a successful first (and more) R experience

1.11.1 SAVING AND THE WORKSPACE OPTION

Save your script and quit RStudio. If you do this, you'll probably be asked if you'd like to save the workspace. Because you saved your script, and because your raw data are safe, there is no need to save the workspace. So we never save it. Well, there are always exceptions, like if you had some really involved analyses that took R much more than a few minutes to complete. Then you could save the workspace. But a better option in these cases is to use the **save**() function. Just to be sure, the two most important things in your R-life are your raw data and your script. Keep those safe and you're sorted.

1.11.2 SOME NICE THINGS ABOUT RSTUDIO

As well as doing the basics very well, RStudio has some very useful, quite advanced features; by using it you are somewhat future-proofing yourself. Hence, from here on, we assume you are using RStudio. As well as organizing the Script Editor, Console, and other panes/windows nicely,

RStudio has other features that can make R-life easier than otherwise. We cover some of these in later chapters, but it's worth exploring the help/cheat sheets for RStudio to see what is on offer. The RStudio team of developers is awesome. RStudio has the following features:

- It works very similarly on Windows PC, Mac, and Linux.
- It allows you to comment or uncomment regions of code.
- It automatically indents code.
- It offers suggestions for completion of code.
- Function help appears while scripting.
- It provides convenient handling of add-on packages.
- It has advanced debugging tools.
- You can easily create documents (reports, presentations) directly from R, using RMarkdown or Sweave.
- It includes advanced version control integration.
- It includes advanced package-building tools.

Appendix 1a   Mini-tutorial solutions

Here's one way to solve the first problem. Before we touch R, let's think through what we need to do. Well, we need some x values and y values to plot against each other. The x values need to go from negative to positive, for example from −10 to 10, and in small enough steps to make a smooth curve. Then we'll need to make the y values, which we'll make equal to x^2. Then we'll make a line plot of y against x. The first two steps we nearly did already in the previous section on playing with R. The plotting is a bit new (with the result shown in Figure 1.10). In an R script, this will be:

```
# Exercise 1
# Plot a graph with x^2 on the y-axis and x on the x-axis.
rm(list=ls())
library(ggplot2)

x <- seq(-10, 10, 0.1)
y <- x^2
qplot(x, y, geom="line")
```

You probably noticed that we're using the plotting function **qplot**(). To use this function, you need to get and load into R the add-on package ***ggplot2***. We use a plotting function called **qplot**() here, rather than **ggplot**(), which we will introduce to you to soon, because it's quick. The 'q' stands for 'quick'.

The solution to the second problem is quite similar. We make an *x* variable, this time from 0 to say 8π, make the *y* variable equal to the sine of the *x* variable, and then use the same plotting command (with the result shown in Figure 1.11):

```
# Exercise 2
# Plot a graph with sine of x on the y-axis and x on the x-axis.
rm(list=ls())
library(ggplot2)

x <- seq(0, 8*pi, 0.1)
y <- sin(x)
qplot(x, y, geom="line")
```

The solution to the third problem is a little different. Here's one way to solve it. First we ask R to make the 1000 random normal deviates and assign them to an object, and then we use the **qplot**() function to plot the histogram (with the result shown in Figure 1.12).

```
# Exercise 3
# Plot a histogram of 1000 random normal deviates.
rm(list=ls())
library(ggplot2)

x <- rnorm(1000)
qplot(x)
```

If you were wondering how the **qplot**() function knows to make a histogram, great. Actually, it guesses, based on it receiving from us only a single variable. That is, **qplot**() thinks, 'I have only one numeric variable to work with, so what is most likely useful? Oh, I know, it's a histogram.'

Appendix 1b   File extensions and operating systems

A note about the three-letter combinations on the end of filenames . . . otherwise known as filename extensions (such as .exe, .csv, .txt). These tell your operating system what application to open a file in if you double-click the file.

When you save a script file, it will get saved with the name you choose followed by the 'filename extension' '.r' or '.R'. Ideally this will mean that when you double-click on a script file, it will automatically open in RStudio. If double-clicked R files are not opening in R, it's rather annoying, and the solution follows . . .

On a Mac and in Windows the default is for hidden file extensions (you don't see them). However, it can be quite useful to see them, especially when they get mangled, by whatever means. We usually ask to see the file extension, since it's not too confusing and is in fact quite useful. If, however, you prefer to not see file extensions, you can still see them if you right-click on the file and select 'Get info' (Mac) or 'Properties' (Windows).

MACINTOSH

To ensure that OSX shows you file extensions, you need to adjust the Finder preferences (Figure 1.13): Finder -> Preferences -> Advanced; tick 'Show all filename extensions'.

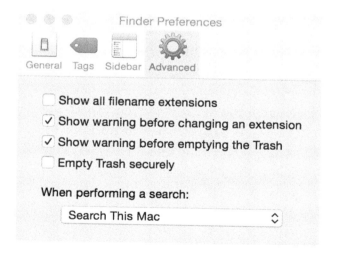

Figure 1.13 The advanced dialogue box of the Finder preferences. Here you can ask OS X to show all file extensions, which you might find useful, or not!

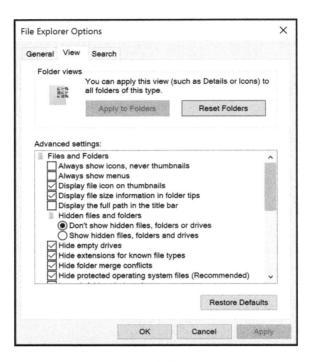

Figure 1.14   The advanced settings tab for File Explorer options in Windows. Here you can ask Windows to show all file extensions, which you might find useful, or not!

WINDOWS

To ensure that Windows presents file extensions, navigate in an Explorer window via the Tools menu to Folder Options: Tools -> Folder Options -> View Tab (Figure 1.14). Ten tick boxes/circles down, you should see an option stating 'Hide extensions for known file types'. Deselect this if it is selected, and Windows should show all file extensions.

FORCING ITEMS TO OPEN WITH RSTUDIO

If double-clicking on a script file doesn't result in it automatically being opened in RStudio, check that your script file has the .r or .R extension (though don't forget that, if you didn't ask, you probably don't see the file extension by default). If it doesn't open, and it doesn't have the correct file extension, add `.r` or `.R` to the end of the filename. Your computer may complain about this: do you really want to change the file type/file

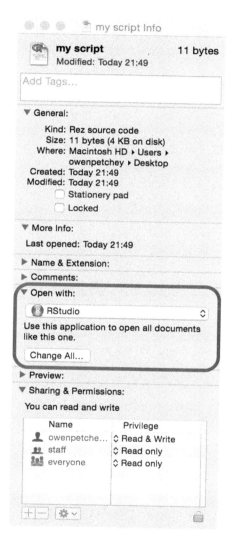

Figure 1.15   The window you get if, on a Mac, you ctrl+click on a file and select *Get Info*. Here you can change the application that opens this file, or even all files of this type (the bits inside the red box).

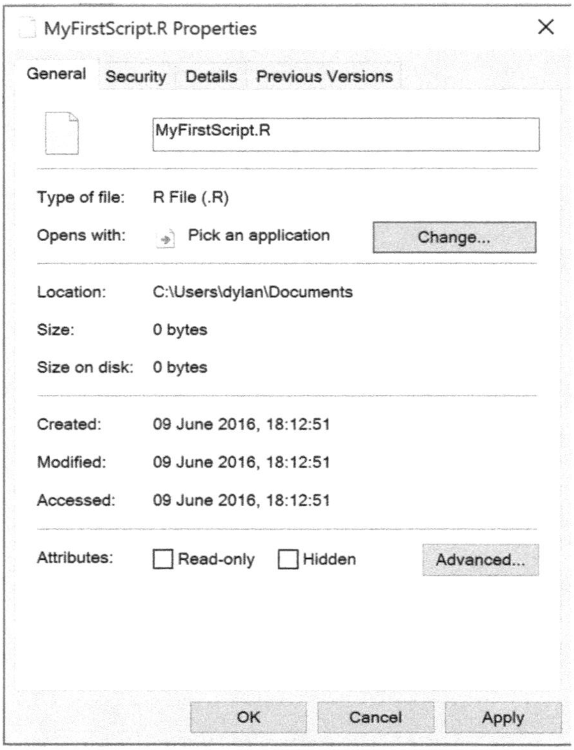

Figure 1.16 The window you get if, in Windows, you right-click on a file and select *Properties*. Here you can change the application that opens this file.

extension? Actually, this is a great check, since you can really upset your computer by changing filename extensions!

If you script still doesn't automatically open in RStudio when double-clicked, right-click on it, go to Get info (Mac, Figure 1.15) or Properties (Windows, Figure 1.16) and look to see which application if any is selected to *open* this file type. Change or set this to RStudio.

2

Getting Your Data into R

By now you must be quite excited that simply by typing some instructions, you can get R to do stuff. You may also be excited about making graphs in R. You have learned many new functions already. . . but there are crucial things to learn about data before we move on.

We can tell . . . you really want to get your data into R's brain. It is empty—R's brain, not yours. This chapter covers four things: how to prepare your data so they will be easy to import into R, and to work with in R; how to import your data into R; and how to check they were correctly imported. We also touch on some troubleshooting skills along the way. And we also provide some methods for how to deal with badly formatted data in the appendix to this chapter. Phew. Data time.

2.1 Getting data ready for R

One of the most frequent stumbling blocks for new students and seasoned researchers using R is actually just getting the data into R. This is really unfortunate, since it is also most people's first experience with R! Let's make this first step very easy.

As in many walks of life, careful preparation is the key to good results. So, hopefully you're reading this before you collected, acquired, or at least

Getting Started with R Second Edition. Andrew Beckerman, Dylan Childs, & Owen Petchey:
Oxford University Press (2017). © Andrew Beckerman, Dylan Childs, & Owen Petchey.
DOI 10.1093/oso/9780198787839.001.0001

typed in the data you'll be working with. There's no disaster if you have already done so, however.

2.1.1 SETTING UP YOUR OWN DATA

R likes data in which there is one observation in a row, and each variable is represented by its own column. Some people call this 'long format'; others call it 'tidy data'. R likes data that look like this in the sense that many of the tools you use in R (the tools are *functions*) work on data that look like this.

Exactly what does it mean to have data with one observation per row and each variable in only one column? Imagine a dataset containing the heights of men and of women. This could be arranged in two columns, one containing the heights of the men, and another the heights of the women (Figure 2.1).

This is *not* what R likes! There are two observations in each row, the values of heights are in two columns, and the gender variable is represented in two columns, even though it is the same variable. The alternative, R-friendly arrangement is to generate one column called 'gender' and a second called 'height' (Figure 2.2). This arrangement has more rows

	A	B
1	Male.height	Female.height
2	138	115
3	161	132
4	183	149
5	136	158
6	183	111
7	186	158
8	174	127
9	167	143
10	191	114
11	147	168

Figure 2.1 An example of not such a good way to arrange data for them to be easily worked with in R. There are two columns that have the same type of information (heights), each row contains two observations.

	A	B
1	Gender	Height
2	Male	138
3	Male	161
4	Male	183
5	Male	136
6	Male	183
7	Male	186
8	Male	174
9	Male	167
10	Male	191
11	Male	147
12	Female	115
13	Female	132
14	Female	149
15	Female	158
16	Female	111
17	Female	158
18	Female	127
19	Female	143
20	Female	114
21	Female	168

Figure 2.2 An example of a good way to arrange data for them to be easily worked with in R. The two columns represent each variable (**gender, height**). Each row contains only one observation, i.e. information about one person.

(twice as many if there are the same numbers of both genders), and this is why it's called 'long format', though a better name might be 'tall format'.

Here is another example gone wrong, and how to fix it. Imagine you re-corded the heights of individuals through time. You could have a separate column for each date at which the height was recorded (Figure 2.3). R doesn't like this, and neither should you.

If anyone gives you data like this, they owe you a <insert preferred beverage>. If you give yourself data like this, do ten push-ups. The better arrangement is to specify Year as a variable—have a column called 'Year' (Figure 2.4). To make good sense, such a dataset would also have columns **Person, Year**, and of course **Height**.

	A	B	C	D	E	F	G	H
1	Person	Height.year1	Height.year2	Height.year3	Height.year4	Height.year5	Height.year6	Height.year7
2	Ellie	138	142	145	150	154	157	162
3	Andrew	161	170	175	182	187	191	191
4	Noah	120	132	140	148	154	159	165
5	Darby	136	145	150	155	159	165	167

Figure 2.3 Ouch! It makes our eyes hurt, and will hurt R too. There are many columns with the same type of information (heights). Each row contains multiple observations.

	A	B	C
1	Person	Year	Height
2	Ellie	1	138
3	Andrew	1	161
4	Noah	1	120
5	Darby	1	136
6	Ellie	2	142
7	Andrew	2	170
8	Noah	2	132
9	Darby	2	145
10	Ellie	3	145
11	Andrew	3	175
12	Noah	3	140
13	Darby	3	150
14	Ellie	4	150
15	Andrew	4	182
16	Noah	4	148
17	Darby	4	155
18	Ellie	5	154
19	Andrew	5	187
20	Noah	5	154
21	Darby	5	159
22	Ellie	6	157
23	Andrew	6	191
24	Noah	6	159
25	Darby	6	165
26	Ellie	7	162
27	Andrew	7	191
28	Noah	7	165
29	Darby	7	167

Figure 2.4 Much better! R will like you if you arrange data like this. There is only one column for each type of information. Each row has only one observation.

An exercise in data preparation

This section tells you how to prepare a data sheet and enter data. You may not have any now, but it's worth reading through to see how it would be done and, specifically, for some conventions for data entry (e.g. about missing values and file formats).

Now that you know how R likes data arranged, make a blank data sheet in Excel (or some other spreadsheet software) and print it out ready for you to fill it in (or keep it on your techno device if you plan to fill it in directly there). Keep column names informative, brief, and simple. Try not to use spaces or special symbols, which R can deal with but may not do so particularly nicely.

Also, if you have categorical variables, such as sex (male, female), use the *actual* category names rather than codes for them (e.g. don't use 1 = male, 2 = female). R can easily deal with variables that contain words, and this will make subsequent tasks in R much more efficient and reliable.

Now . . . in between making this sheet and coming back to R, you will have done some science and filled in your data sheet. Rest assured you can continue using the book. We will provide you with a dataset to carry on.

Top Tip 1. As you get to the data entry point of your work, *remember*, all cells in the data sheet should have an entry, even if it is something saying 'no observation was made'. The expected placeholder for missing data in R is NA. Other options are possible, but be careful. A blank cell is ambiguous: perhaps you just forgot to make the observation, or perhaps it really couldn't be made. Good practice is to look over the data sheet every hour or so, checking for and adding entries in empty cells.

Once you've entered the data on the data sheet, type your data into a spreadsheet program. Once complete, print your entered data onto paper, and check that this copy of your data matches the data on your original data sheets. Correct any mistakes you made when you typed your data in.

Top Tip 2. We don't advocate saving your file as an .xls, .xlsx, .oo, or .numbers file. Instead, we actively argue for using a 'comma-separated values' file (a .csv file). A .csv file is easily transported with a low memory and small size footprint among computing platforms. In Excel, Open Office, or Numbers, after you click Save As . . . you can change the format of the file to 'comma-separated values', then press Save. Excel might then, or when you close the Excel file, ask if you're sure you'd like to save your data in this format. Yes, you are sure!

At this point in our workflow, you have your original paper data sheets, a digital copy of the data in a .csv file, and a printed copy of the data from the `.csv` file. One of the remarkable things about R is that once a copy of your 'raw data' is established, the use of R for data visualization and analysis will never require you to alter the original file (unless you collect more data!). Therefore keep it very safe! In many statistical and graphing programs, your data spreadsheet has columns added to it, or you manipulate columns or rows of the spreadsheet before doing some analysis or graphic. With R, your original data file can and must always remain unmanipulated. If any adjustments do occur, they occur in an R copy only and only via your explicit instructions harboured in the script.

2.1.2 SOMEONE ELSE'S DATA?

And what if you already have your data, and they're not in the arrangement R likes? Or perhaps your data came from someone else and you didn't have a chance to tell them how R likes data arranged? Or your data were recorded by a machine, and you therefore didn't have much choice about how they were recorded? (Fortunately, most machines seem to know how R likes data to be arranged, and therefore adhere to the 'one observation per row and a variable per column' rule.) One thing you can do is send it back, saying how you need it! When you think such a request might not go down too well, you're going to have to do the work yourself.

You can rearrange the data in Excel, though this can cause errors, can be awfully time-consuming, and is tedious and boring for big datasets. The alternative is to let R do the hard work: let R rearrange the data. We provide details on how to do this in the appendix at the end of this chapter.

2.2 Getting your data into R

Brilliant . . . you now have your beautiful data ready for R. If you just happened to miss our emphasis above, it should be a long/tall-format `.csv` file. But how do you get the data in this `.csv` file into R? How do you go from making a dataset in a `.csv` file to having a working copy in R's brain to use for your analysis?

First things first ... you will need the datasets that we use in this book. You can get these from http://www.r4all.org/the-book/datasets. Don't forget to unzip them and put them somewhere nice! Part of keeping your data safe is putting it somewhere safe. But where? We like to be organized, and we want you to be organized too.

Make a new folder in the main location/folder in which you store your research information—perhaps you have a Projects folder in MyDocuments (old PC) or Documents (new PC and Mac)—create a folder inside this location. Give the folder an informative name—perhaps 'MyFirstAnalysis'. It should really go without saying that you should be able to find this folder, i.e. you should be able to navigate to it within Explorer (Windows) or Finder (Mac), or whatever the equivalent is in your favourite Linux.

Now, inside this folder you just made, make another folder called Analyses (i.e. 'MyFirstAnalysis/analyses'). And then another: Datasets. At this point, you should move the datasets for the book to the Datasets folder you just made.

If you're working on a full project, you might perhaps make another folder called Manuscript and perhaps another one called Important PDFs. We think you might be getting the point: use folders to organize separate parts of your project. Your file of instructions for R—your *script* file—will go in the Analyses folder. We'll build one soon. Be patient. ☺

2.2.1 IMPORTING DATA, PREPARATION

For R to import data (i.e. to get your data into R's brain), you need to tell R the location of your data. The location, or address, for information on a computer is known as the *path*, and these are often rather long, difficult to remember, and even harder to type accurately. And if you make one little mistake while typing your path, R will be lost. So never type your path. Let RStudio do the hard work.

First, open Rstudio and open a new script file. Put the preliminary information (remember this from the previous chapter?) at the start of your script— some annotation, the libraries, and the brain clearer. Go ahead and save this. Perhaps call it `DataImportExample.R`?

We don't expect you to use your own data for this. In the next few sections and also the next chapter, we will work with the `compensa-tion.csv` dataset.

There are at least four really easy ways to get R to import the data, without having to do anything tricky like type the path to your data file.

2.2.2 METHOD 1: THE IMPORT DATASET TOOL

Rstudio provides menu-based import functionality via the `Import Dataset` tab in the upper right pane of RStudio. This will allow you to navigate to your data file, select it, and click the *Open* button. When you attempt to use this for the first time, RStudio may ask you to install or update some packages; just say yes! These are packages that make data import faster and simpler.

Given our focus on .csv files, select *From CSV* as your option. There are clearly others. The dialogue box will open revealing a very handy tool (Figure 2.5).

First, click the *Browse* button and navigate to your data file in the window that pops up. Click *Open* and you will see how R is interpreting your data (it is not yet into R, you're just getting a preview). At this point, you could fiddle with some of the import options in the lower left part of the window (e.g., the type of delimiter / separator). Often you don't have to change anything, and can just click the *Import* button.

Before you do anything more, copy the first two lines in the code preview box (example shown in Fig 2.6). Then, click *Import*. RStudio will run all of the code in the preview box. This imports the data using the **read_csv()** function from the readr library. It also opens a window showing the data. Close that window - you don't need it open.

Now, paste the two lines of code into your script. Having done this, you don't have to use the `Import Dataset` tool again. You just run these lines of code and your data are read in (assuming you don't move the data . . .). Your script now contains the information about which dataset is being used and where it is on your computer. You run this code each time you work with the script.

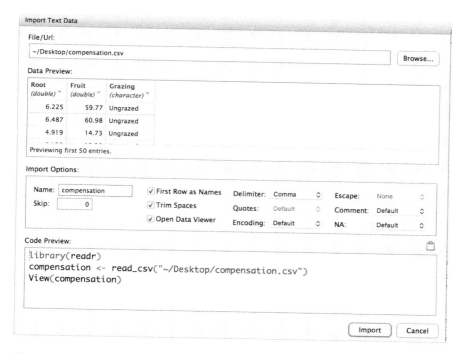

Figure 2.5   The *Import Dataset* tab produces this dialogue box, with details about the format, delimiter (e.g. comma), and values being used for NA (the na.string).

```
Code Preview:
library(readr)
compensation <- read_csv("~/Desktop/compensation.csv")
```

Figure 2.6   An example of the two lines of instructions produced by the *Import Dataset* tool in Rstudio.

Top Tip. Do not rely on using the `Import Dataset` tool each time you work on an analysis. This would be inefficient but, more importantly, not pasting in the R code snippet leaves you with an incomplete script, and if you keep reading your data in via `Import Dataset` you may end up working with the wrong file.

2.2.3 METHOD 2: THE **file.choose**() FUNCTION

If you don't use Rstudio, or even if you do, you can use the **file.choose**() function to get your path and filename into R without typing it. In the

```
> file.choose()
[1] "/Users/owenpetchey/work/0 research/5.published/Dilpetus/Dileptus data/dileptus
expt data.csv"
```

Figure 2.7 An example of the instruction produced when you use the **file.choose**() function to find a data file.

Console, type **file.choose**() and press enter. A dialogue box will open, implicitly inviting you to navigate to your data file, to select it, and to press open. In the Console, you will then see the path and filename. Figure 2.7 gives an example of the Console after you've done this.

Select the double quotes and everything inside them (and not the [1]), copy this, and paste it into your script. There, you have your path and filename. Easy, eh! Now you need to give this information to the function that reads in the data file . . . this function is called **read.csv**() because, no surprises, it reads a csv file into R. So put the path and filename inside the brackets of **read.csv**() and use the assignment arrow (<-) to save the imported data into an object with a name of your choice. In the example below, this object is called compensation. Remember that this instruction should now be in your script file, and not typed directly into the Console. As above, for this script/project, and via this method, you don't have to use the **file.choose**() function again.

2.2.4 METHOD 3: DANCING WITH THE IMPORT DEVIL

There is another solution. It DOES require that you know where you are working within the hierarchy of your computer. In RStudio, you can set the working directory for a session of coding: Session -> Set Working Directory -> Choose Directory. This will point RStudio's eyes *directly* at a folder in a specific location, the folder where the data are.

Once RStudio is looking there, the use of **read.csv**() becomes 'path free'; you need only supply the full name of the dataset, in quotation marks, for example compensation <- **read.csv**("compensation.csv"). The advantage of doing things this way is that if you move your work, or collaborate with someone who keeps the project in a different location on their machine, you don't have to update your file paths. Just set the working directory and away you go.

As we note above, this can be dicey for some of you (and some of us). But if you organized your files in folders that are sensibly named and located, there will be no trouble. In fact, after a few uses of *Import Dataset* via RStudio, you will know more about the path than ever before.

2.2.5 METHOD 4: PUT YOUR DATA IN THE SAME PLACE AS YOUR SCRIPT

If you put your data and script in the same folder (which may or may not be such a good idea) and you double-click on the script, and this causes RStudio to start (because it wasn't already running), then the default place RStudio will look for data is in the folder with the script. In this case, you don't need the path at all. It's a bit risky to rely on this, since it only works if RStudio was *not* already running. Also, you often don't want your data in the same folder as your script. The solution is then to use relative paths, but that is beyond what we want to go into here. After you get comfortable with important data and paths, you can maybe take a moment to Google for 'using relative paths R' if you're interested.

2.3 Checking that your data are your data

A very sensible next step is to make sure the data you just asked R to commit to its brain are actually the data you wanted. Never trust anyone, let alone yourself, or R. Some basic things to check are:

- The correct number of rows are in R.
- The correct names and number of variables are in R.
- The variable are of the correct type (e.g. R recognizes numeric variables to be numeric).
- Variables describing types of things (e.g. gender) have the correct number of categories (levels).

If you have imported the compensation data, you should have an object called compensation. A few very sensible functions to commit to your memory are presented below, and we cover what they produce. You should feel free to add these to your script:

```
names(compensation)

## [1] "Root"     "Fruit"     "Grazing"

head(compensation)

##      Root Fruit   Grazing
## 1  6.225 59.77 Ungrazed
## 2  6.487 60.98 Ungrazed
## 3  4.919 14.73 Ungrazed
## 4  5.130 19.28 Ungrazed
## 5  5.417 34.25 Ungrazed
## 6  5.359 35.53 Ungrazed

dim(compensation)

## [1] 40    3

str(compensation)

## 'data.frame':      40 obs. of   3 variables:
## $ Root    : num   6.22 6.49 4.92 5.13 5.42 ...
## $ Fruit   : num   59.8 61 14.7 19.3 34.2 ...
## $ Grazing: Factor w/ 2 levels "Grazed",
                  "Ungrazed": 2 2 2 2 2 2 2 2 2 ...
```

Not surprisingly, **names**() tells you the names you assigned each column (i.e. your variable names, the column names in your .csv files, which you typed into Excel or wherever). **head**() returns the first six rows of the dataset (guess what **tail**() does?). **dim**() tells you the numbers of rows and columns—the *dimension*—of the dataset. Finally, **str**() returns the structure of the dataset, combining nearly all of the previous functions into one handy function.

The function **str**() returns several pieces of information about objects. For a *data frame* (spreadsheet-like dataset), it returns in its first line a statement saying that you have a data frame type of object, and the number of observations (rows) and variables (columns). Then there is a line for each of your variables, giving the variable name, variable type (numeric, factors, or integers, for example), and the first few values of that variable. Thus, str() it is an outstanding way to ensure that the data you have imported are what you intended to import, and to remind yourself about features of your data.

Of course, you could also just type the name of the dataset— compensation. This can be pretty fun/infuriating if you forget that there are 10,000+ rows in your dataset.

2.3.1 YOUR FIRST INTERACTION WITH *dplyr*

In the previous chapter, we had you install the package *dplyr*. If you have been working along, and you've run the script examples from Chapter 1, *dplyr* should be in and ready to use. Two functions from *dplyr* are also useful for looking at the data you've just imported. One, **glimpse()**, provides a horizontal view of the data. The second, **tbl_df()**, provides a vertical view. Both show the column names and the type of data.

We will emphasize this over and over again as we move on in the book: the functions in *dplyr* and *ggplot2*, and in fact all packages from Hadley Wickham, all have the SAME first argument—the data frame. We'll introduce more *dplyr* functions in Chapter 3, all of which take the data frame as their first argument.

Let's see what **glimpse()** and **tbl_df()** do:

```
# dplyr viewing of data
library(dplyr)

#glimpse and tbl_df
glimpse(compensation)

## Observations: 40
## Variables: 3
## $ Root    (dbl) 6.225, 6.487, 4.919, 5.130, 5.417, 5.35...
## $ Fruit   (dbl) 59.77, 60.98, 14.73, 19.28, 34.25, 35.5...
## $ Grazing (fctr) Ungrazed, Ungrazed, Ungrazed, Ungrazed...

tbl_df(compensation)

## Source: local data frame [40 x 3]
##
##      Root Fruit  Grazing
```

```
##      (dbl)  (dbl)    (fctr)
## 1   6.225  59.77  Ungrazed
## 2   6.487  60.98  Ungrazed
## 3   4.919  14.73  Ungrazed
## 4   5.130  19.28  Ungrazed
## 5   5.417  34.25  Ungrazed
## 6   5.359  35.53  Ungrazed
## 7   7.614  87.73  Ungrazed
## 8   6.352  63.21  Ungrazed
## 9   4.975  24.25  Ungrazed
## 10  6.930  64.34  Ungrazed
## ..    ...     ...       ...
```

Nice. We get two simple summaries of the data inside `compensation`.
We are going to use **glimpse**() from now on (mostly).

2.4 Basic troubleshooting while importing data

At this point you know how to create a `.csv` file of your data, import these
data, and make sure the data in R are the data you wanted, using several
functions. Of course, along the way, bad things will happen. The magic will
fail. Something just won't work. Here are a few that we find happening to
new R people all the time.

Problem. *My imported data has more columns and/or rows than it
should! (and there are many, many NA's!)* This likely to be caused by Excel,
which has saved some extra columns and/or rows (goodness knows why!).
To see if this is the problem, open your data file (`.csv` file) using Notepad
(Windows) or TextEdit (Mac). If you see any extra commas at the end
of rows, or lines of commas at the bottom of your data, this is the problem. To fix this, open your `.csv` file in Excel, select only the data you
want, copy this, and paste this into a new Excel file. Save this new file
as before (and delete the old one). It should not have the extra rows/
columns.

Problem. *There's only one column in the imported data, but definitely
more than one in my data file!* This is probably caused by your file not being
'comma separated', but R expecting it to be. Most often this happens when

Excel decides it wants to save a .csv file with semicolon (;) separation, instead of comma! There are several options. On the Excel side, try and ensure that Excel is using commas for .csv files (though this is sometimes easier said than done). On the R side, try using the *Import Dataset* facility in Rstudio; during the process, you can see the raw data and the imported data and look at the raw data to see what the separator is, and then select this in the dialogue box (see Figure 2.5).

Problem. *R gives this error:*

```
## Warning in file(file, "rt"): cannot open file 'blah.csv': No
such file or directory
## Error in file(file, "rt"): cannot open the connection
```

Read the first line of that error message: 'No such file or directory'. Did you type, rather than copy and paste, the path or filename? If so, you probably made a mistake. Or perhaps you moved or renamed the file? This is a very common error associated with trying to find your data. Use the *Import Dataset* tool in Rstudio, or the **file.choose**() function again, and you will fix this error.

Problem: *My dataset contains a column of dates, but I think R is not recognizing or treating them as dates!* You are justifiably concerned, because R won't, by default, read in your dates as real dates. Look at the example in the appendix about tidying data in this chapter, for how to get R to recognize a variable as containing dates.

2.5 Summing up

You just learned how to prepare your data for R, how to import your data, how to check they are properly in R, and quite a bit about what to do if things go wrong. These tasks are the basic first steps in any process of quantitative problem solving, and they are the foundation. They are not optional steps. Forget these, get these wrong, and everything else will be wrong also. So focus on having rock-solid data in R.

Appendix Advanced activity: dealing with untidy data

Health warning: this section is not necessary if you're working through the book for the first time. It's also a bit complex, so think about skipping it, saving it for when you're happy to take a bit of a challenge (great if that is now, however . . . go for it!).

This section uses several new packages that you will need to install first (e.g. download from CRAN). It uses many functions we have not introduced, and we don't introduce many in depth. If you prepare your data well, this section is unnecessary. But you will encounter messy data. So, as we said, feel free to come back to it.

WHAT ARE UNTIDY DATA?

There are lots of ways data can be untidy, and it's not appropriate (or possible) to deal with them all here. So we show a somewhat typical example in which observations of the same thing (here bacterial density) have been recorded on multiple dates, and the observation from each date is placed in a separate column. Rows of the data correspond to experimental units (defined here by the protist species consuming the bacteria, and the environmental temperature) (another variable, **Bottle**, is a unique identifier for the experimental unit). Let's import these data, check them, and then tidy them:

```
nasty.format <- read.csv("nasty format.csv")

str(nasty.format)

## 'data.frame':    37 obs. of  11 variables:
## $ Species : Factor w/ 5 levels "",
##     "Colpidium",..: 3 3 3 3 3 3 3 3 3 5 ...
## $ Bottle  : Factor w/ 37 levels "","10-C.s","11-
##     C.s",..: 35 36 37 11 12 13 23 24 25 5 ...
## $ Temp    : int  22 22 22 20 20 20 15 15 15 22 ...
## $ X1.2.13 : num  100 62.5 75 75 50 87.5 75 50 75 37.5 ...
## $ X2.2.13 : num  58.8 71.3 72.5 73.8 NA NA NA
##     NA NA 52.5 ...
## $ X3.2.13 : num  67.5 67.5 62.3 76.3 81.3 62.5 90
```

```
         78.8 78.3 23.8 ...
##  $ X4.2.13 : num   6.8 7.9 7.9 31.3 32.5 28.8 72.5
         92.5 77.5 1.25 ...
##  $ X6.2.13 : num   0.93 0.9 0.88 3.12 3.75 ...
##  $ X8.2.13 : num   0.39 0.36 0.25 1.01 1.06 1 67.5
         72.5 60 0.96 ...
##  $ X10.2.13: num   0.19 0.16 0.23 0.56 0.49 0.41
         37.5 52.5 60 0.33 ...
##  $ X12.2.13: num   0.46 0.34 0.31 0.5 0.38 ...
```

First things first. The data frame in R has 37 observations and 11 variables. The experiment involved only 36 experimental units, however. Looking at the data in R, by clicking on the `nasty.format` text in the Rstudio Environment pane, and scrolling down in the displayed data, shows there is a 37th row containing no data. This is Excel trying to trip us up (actually, it's usually caused by something odd *we* did in Excel). Let's remove that row. We do this by looking at the row and seeing what is unique about it, compared with all the others. Importantly, it lacks an entry in the **Bottle** variable. So we tell R to keep only rows that have an entry in the **Bottle** variable, using the **filter()** function in the *dplyr* package (more about this add-on package, and **filter()**, in the next chapter):

```
library(dplyr)
nasty.format <- filter(nasty.format, Bottle != "")
glimpse(nasty.format)

## Observations: 36
## Variables: 11
## $ Species  (fctr) P.caudatum, P.caudatum, P.caudatum, P...
## $ Bottle   (fctr) 7-P.c, 8-P.c, 9-P.c, 22-P.c, 23-P.c, ...
## $ Temp     (int) 22, 22, 22, 20, 20, 20, 15, 15, 15, 22...
## $ X1.2.13  (dbl) 100.0, 62.5, 75.0, 75.0, 50.0, 87.5, 7...
## $ X2.2.13  (dbl) 58.8, 71.3, 72.5, 73.8, NA, NA, NA, NA...
## $ X3.2.13  (dbl) 67.5, 67.5, 62.3, 76.3, 81.3, 62.5, 90...
## $ X4.2.13  (dbl) 6.80, 7.90, 7.90, 31.30, 32.50, 28.80,...
## $ X6.2.13  (dbl) 0.93, 0.90, 0.88, 3.12, 3.75, 3.12, 10...
## $ X8.2.13  (dbl) 0.39, 0.36, 0.25, 1.01, 1.06, 1.00, 67...
## $ X10.2.13 (dbl) 0.19, 0.16, 0.23, 0.56, 0.49, 0.41, 37...
## $ X12.2.13 (dbl) 0.46, 0.34, 0.31, 0.50, 0.38, 0.46, 41...
```

TIDYING WITH GATHER()

Now we make the data tidy. We need a new variable that contains the date on which observations were made, and a new variable containing the observations of bacterial abundance that are currently stored in columns 4 to 11. And then we need to move the data into these new columns. This is all very straightforward, thanks to the **gather**() function in the *tidyr* package. Use it like this:

```
library(tidyr)
tidy_data <- gather(nasty.format, Date, Abundance, 4:11)
```

This first argument of **gather** is the data frame to work on: nasty.format. The second is the name of the new variable that will contain the dates (we call it **Date**). The third is the name of the new variable that will contain the bacterial abundances (we call it **Abundance**). The fourth argument is the location in the nasty.format data frame of the observations of bacterial abundance that are to be put into the new **Abundance** variable. Use **str**() to see if it worked:

```
glimpse(tidy_data)

## Observations: 288
## Variables: 5
## $ Species   (fctr) P.caudatum, P.caudatum, P.caudatum, ...
## $ Bottle    (fctr) 7-P.c, 8-P.c, 9-P.c, 22-P.c,
    23-P.c, ...
## $ Temp      (int) 22, 22, 22, 20, 20, 20, 15, 15, 15, 2...
## $ Date      (chr) "X1.2.13", "X1.2.13", "X1.2.13", "X1....
## $ Abundance (dbl) 100.0, 62.5, 75.0, 75.0, 50.0, 87.5, ...
```

Yes, super! We have 288 observations (36 experimental units (bottles), each observed on eight dates). We have the new **Date** variable, a `factor`-type variable with eight levels, and the new **Abundance** variable, which is numeric. (Give yourself a pat on the back if you wonder why there are still 37 levels in the **Bottle** variable; though it isn't too much of a problem to correct at present.)

CLEANING THE DATES

The data are now officially *tidy*. The **Date** variable still needs a bit of work, however. We need to remove the X at the beginning, and make R recognize that these are dates (i.e. change the variable type from Factor to a date-type variable.). First we remove the X, using the **sub_str()** function in the stringr add-on package:

```
library(stringr)
tidy_data <- mutate(tidy_data, Date = substr(Date, 2, 20))
```

We tell the **sub_str()** function the variable to work on, and the character in that variable at which to start keeping characters (i.e. keep from the second character on). The third variable we have made 20, which is past the end of the date, so we don't discard any later characters. We do all this within the **mutate()** function of the *dplyr* add-on package, which provides a neat way of changing (or adding) variables in a data frame (much more about this in the next chapter).

Now we need to change the variable into one that R recognizes as containing dates. Technically, we are going to *parse* the information in the date variable so that R regards it as a date. This will allow us to use the Date variable as a continuous variable, for example to make a graph with the date on the *x*-axis, or even to calculate the number of days, months, or years between two or more dates.

We will use a function in the *lubridate* package. This package contains the functions **ymd()**, **ydm()**, **dym()**, **dmy()**, **myd()**, and **mdy()**, among others. Which of these functions should we use for our data? Take a look at the date values:

```
unique(tidy_data$Date)
```

```
## [1] "1.2.13"  "2.2.13"  "3.2.13"  "4.2.13"  "6.2.13"
## [6] "8.2.13"  "10.2.13" "12.2.13"
```

(Note that we've used a bit of classic R here, the dollar sign, to refer to a variable inside a data frame. Sometimes the classic way is still very useful!)

Not too much sleuthing shows that our date has the format day.month.year. So we use the function **dmy()**. This is a quite intelligent function, able to deal with different separators (ours is a dot but others work), and dates that include leading zeros (ours does):

```
library(lubridate)
tidy_data <- mutate(tidy_data, Date = dmy(Date))
```

And look at the type of variable the dates are now:

```
glimpse(tidy_data)

## Observations: 288
## Variables: 5
## $ Species    (fctr) P.caudatum, P.caudatum, P.caudatum, ...
## $ Bottle     (fctr) 7-P.c, 8-P.c, 9-P.c, 22-P.c,
##   23-P.c,...
## $ Temp       (int) 22, 22, 22, 20, 20, 20, 15, 15, 15, 2...
## $ Date       (date) 2013-02-01, 2013-02-01, 2013-
##   02-01, ...
## $ Abundance  (dbl) 100.0, 62.5, 75.0, 75.0, 50.0, 87.5, ...
```

It is a POSIXct-type variable if you use str() or it is a date variable
if you use glimpse(). Look POSIXct up in Google if you care to know
exactly what it means. It's enough, however, to know that this means that
R knows this variable contains dates.

NOW SEE WHAT WE CAN DO! . . .

We just tidied our data, and made the dates into dates. Maybe that seemed
a lot of effort, so we'll take a moment to demonstrate one example of what
this allows us to do. Imagine we want to view the dynamics of bacterial
abundance in each bottle. We can now do this with very little code:

```
library(ggplot2)
ggplot(data = tidy_data, aes(x=Date, y=Abundance)) +
  geom_point() +
  facet_wrap(~Bottle)
```

First, if you're not familiar with making graphs using ggplot(), go to
Chapter 4, where we explain everything. We tell ggplot() the data frame
to look in for variables, the *x*- and *y*-variables, to plot points, and then
ask for on graph (i.e. facet) for each of the bottles. The resulting panel of
graphs (Figure 2.8) has correct dates on the *x*-axis, and is really useful for
data exploration, and even publication after some polishing (e.g. making
the *x*-axis tick labels legible). A panel of graphs like this would have been

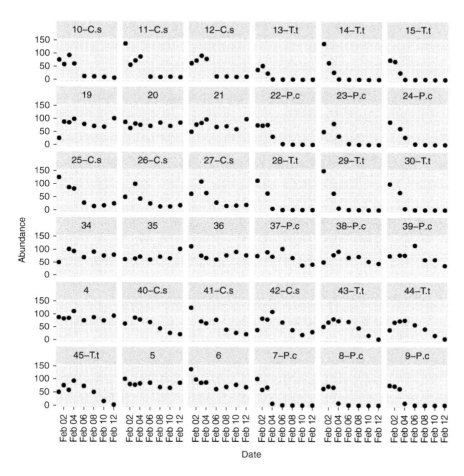

Figure 2.8 Figures like this are a doddle if one has tidy data (just three short lines of code was used to make the graph).

considerably more hassle to make if the data were not tidy, i.e. if the data were left as they were in the data file, with one variable for each date.

Why not rearrange the data in Excel (or any spreadsheet program)?

1. We have better things to do with our time, like poking ourselves in the eye with a red-hot skewer.
2. We will make mistakes, eventually, and probably won't know.
3. We'll have to do it all over again if we change the dataset.

As we mentioned, there are many ways in which data can be untidy and messy. Fortunately, there are lots of nice functions for dealing with untidy and messy data. In the ***base***, ***tidyr***, and ***dplyr*** packages there are:

- **spread**(): does the opposite of **gather**. Useful for preparing data for multivariate methods.
- **separate**(): separates information present in one column into multiple new columns.
- **unite**(): puts information from several columns into one column.
- **rename**(): renames the columns of your data.
- **rbind**(): puts two datasets with exactly the same columns together (i.e. it binds rows together).
- **cbind**(): puts two datasets with exactly the same rows together (i.e. it binds columns together). (often better to use the next function . . .)
- **join**(): a suite of functions, such as **full_join**(), which allows joining together two datasets with one or more columns in common. (Same as the **merge**() function in the ***base*** package.)

You will probably find that tidying data, especially data that you didn't prepare, can take a lot of time and effort. However, it's investment well made, as having tidy and clean data makes everything following ten times easier.

We've used functions from several add-on packages: ***tidyr***, ***stringr***, ***lubridate***, ***dplyr***, and ***ggplot2***. It would be well worth your while to review these, and make notes about which functions we used from each of these packages, why we used them, and the arguments we used.

Data Management, Manipulation, and Exploration with *dplyr*

Exploring, manipulating, and graphing your data are fundamental parts of data analysis. In fact, we put huge emphasis on spending time developing familiarity with the data you've collected, culminating in a figure that reflects the questions you set out to ask when you actually collected the data. Our workflow starts with making a picture. Before you do any statistics. Because, surely, what you would love to see is that the patterns you *expected* to see in your data are actually there.

There are two fundamental toolsets for this initial effort: manipulation tools and graphing tools. In this chapter, we introduce the ***dplyr*** package to do various common data manipulations. This is extremely important learning, as we almost always need to do some data manipulation, such as subsetting or calculation of mean ± SE. You should do all your data manipulation in R (and none anywhere else), and this chapter gives you the tools to do this. It enables your entire workflow to be contained, as far as is humanly possible, in one place: your R script.

This chapter is definitely one you should work along with, building a script as you go. If you are not yet comfortable with libraries, importing data, and annotating a script, you will be by the end. Review the tips and

Getting Started with R Second Edition. Andrew Beckerman, Dylan Childs, & Owen Petchey:
Oxford University Press (2017). © Andrew Beckerman, Dylan Childs, & Owen Petchey.
DOI 10.1093/oso/9780198787839.001.0001

tricks from the previous chapters about sending data from the script to R. As you go through this chapter, build a script, with copious annotations. Type the commands into the script, and submit them to the Console. Do not type them direct into the Console! Be strict with yourself about both these things. Building good habits now is really important.

We'll be working with the compensation data we imported in the previous chapter, so you should be ready to go. (The compensation data are available at `http://www.r4all.org/the-book/datasets`). You should have **dplyr** and **ggplot** installed already, so you can set up a script with some initial annotation, the **library**() function, and the brain-clearing **rm**(list = **ls**()). This should be followed by use of **read.csv**() to import the `compensation.csv` dataset.

In the Preface we described why we are teaching you to use **dplyr** and its functions, rather than the classic ways of managing and manipulating data. Nevertheless, you may be curious about the classic ways, perhaps because you have some experience with them. In an appendix to this chapter, we briefly make some comparisons between the classic ways and the newer **dplyr** ways.

3.1 Summary statistics for each variable

3.1.1 THE COMPENSATION DATA

Just so you have a frame of reference going forward, the compensation data are about the production of fruit (apples, kg) on rootstocks of different widths (mm; the tops are grafted onto rootstocks). Furthermore, some trees are in parts of the orchard that allow grazing by cattle, and others are in parts free from grazing. Grazing may reduce the amount of grass, which might compete with the apple trees for resources.

3.1.2 THE **summary**() FUNCTION

We covered a few handy tools in the previous chapter for looking at pieces of your data. While **names**(), **head**(), **dim**(), **str**(), **tbl_df**(), and **glimpse**() are good for telling you what your data look like (i.e. correct number of rows and columns), they don't give much information about the particular values in each variable, or summary statistics of these. To get this information, use the **summary**() function on your data frame. Go ahead and apply the **summary**() function to these data:

```
compensation <- read.csv("compensation.csv")
glimpse(compensation) # just checkin'

# get summary statistics for the compensation variables
summary(compensation)
```

```
##        Root              Fruit            Grazing
##   Min.   : 4.426    Min.   : 14.73    Grazed  :20
##   1st Qu.: 6.083    1st Qu.: 41.15    Ungrazed:20
##   Median : 7.123    Median : 60.88
##   Mean   : 7.181    Mean   : 59.41
##   3rd Qu.: 8.510    3rd Qu.: 76.19
##   Max.   :10.253    Max.   :116.05
```

The **summary**() function in R gives us the median, mean, interquartile range, minimum, and maximum for all numeric columns (variables), and the 'levels' and sample size for each level of all categorical columns (variables). It's worth looking carefully at these summary statistics. They can tell you if there are unexpectedly extreme, implausible, or even impossible values—a good first pass at your data. Now let's continue to use these data to learn how *dplyr* can explore, subset, and manipulate data.

3.2 *dplyr* verbs

We are now going to introduce to you five 'verbs' that are also functions in *dplyr*—a package focused on manipulating data. The verbs are **select**(), **slice**(), **filter**(), **arrange**(), and **mutate**(). **select**() is for selecting columns, and **slice**() is for selecting rows. **filter**() is for selecting subsets of rows, conditional upon values in a column. **arrange**() is for sorting rows and **mutate**() is for creating new variables in the data frame. These are core functions for core activities centred on grabbing pieces of, or subsetting, your data (e.g. **select**(), **slice**(), **filter**()) transforming variables (e.g. **mutate**()), or sorting (e.g. **arrange**()). Of course, there are more . . . a quick Google search on 'cheatsheet ***dplyr***' or following the RStudio -> Help Menu -> Cheatsheets trail will take you to lovely places.

The key to using ***dplyr*** is to remember that the first argument to ALL ***dplyr*** functions is the data frame. You might try saying this 25 times. The first argument for ***dplyr*** functions is . . .

3.3 Subsetting

 Three verbs form the core of subsetting data: they get columns, rows, or subsets of rows.

3.3.1 select()

select() grabs columns. Of course, it helps to know the name of the columns, so if you need to, use **names**(compensation) first. Here is how we can use it to get the **Fruit** column (we have truncated the output; you will see more rows than we have printed):

```
select(compensation, Fruit)  # use the Fruit column


## Source: local data frame [40 x 1]
##
##     Fruit
##     (dbl)
## 1   59.77
## 2   60.98
## 3   14.73
## 4   19.28
## 5   34.25
## 6   35.53
## 7   87.73
## 8   63.21
## 9   24.25
## 10  64.34
## ..    ...
```

Note: If you get an error Error: could not find function "select" then you have either not put **library**(dplyr) at the top of your script, or have not run that line of code.

So, let's work through some details about how *dplyr* works. First, you can see that it is quite easy to use. If we want a column, we tell *dplyr* which dataset to look in, and which column to grab. Super.

The astute observer will recognize something else very interesting. select(), as a *dplyr* verb, uses a data frame *and* returns a data frame. If you scroll up to see the top of the output, you will see the column name 'Fruit'. You have asked for part of a data frame, and you get one back—in this case a one-column data frame. Not all R functions act like this. The appendix to this chapter provides some detail on base (classic) R functions that do similar things to *dplyr* functions, but can return different types of object.

Finally, you may also notice that **select**() seems to do one thing only. This is totally true. All *dplyr* functions do one thing, and one thing *very* fast and *very* effectively.

select() can also be used to select all columns *except* one. For example, if we wanted to leave out the Root column, leaving only the Fruit and Grazing columns:

```
select(compensation, -Root)   # that is a minus sign


## Source: local data frame [40 x 2]
##
##      Fruit  Grazing
##      (dbl)    (fctr)
## 1   59.77 Ungrazed
## 2   60.98 Ungrazed
## 3   14.73 Ungrazed
## 4   19.28 Ungrazed
## 5   34.25 Ungrazed
## 6   35.53 Ungrazed
## 7   87.73 Ungrazed
## 8   63.21 Ungrazed
## 9   24.25 Ungrazed
## 10 64.34 Ungrazed
## ..   ...       ...
```

3.3.2 slice()

slice() grabs rows. It works by returning specific row numbers you ask for. You can ask for one row, a sequence, or a discontinuous set. For example, to get the second row, we use

```
slice(compensation, 2)
```

```
##    Root Fruit  Grazing
## 1 6.487 60.98 Ungrazed
```

If we want the second to the tenth, we can invoke the : to generate the sequence:

```
slice(compensation, 2:10)
```

```
##    Root Fruit  Grazing
## 1 6.487 60.98 Ungrazed
## 2 4.919 14.73 Ungrazed
## 3 5.130 19.28 Ungrazed
## 4 5.417 34.25 Ungrazed
## 5 5.359 35.53 Ungrazed
## 6 7.614 87.73 Ungrazed
## 7 6.352 63.21 Ungrazed
## 8 4.975 24.25 Ungrazed
## 9 6.930 64.34 Ungrazed
```

And discontinuous sets are easy, but we need to *collect* the row numbers using another helper function in R called c():

```
slice(compensation, c(2, 3, 10))
```

```
##    Root Fruit  Grazing
## 1 6.487 60.98 Ungrazed
## 2 4.919 14.73 Ungrazed
## 3 6.930 64.34 Ungrazed
```

One thing you may notice about slice() is that it also returns a data frame, but it does not return the row number identity found in the original data. You have new, continuous row numbers. Just be aware.

3.3.3 filter()

filter() is super-powerful subsetting. It requires some basic knowledge of logical operators and boolean operators in R. Let's first work through those, and then learn how to apply them via filter().

Logical operators and booleans

R has a complete set of logical operators. In Table 3.1, we provide some insight into common logical and boolean operators, with examples of their use in **filter**().

One of the tricks you can use to understand how this works can be seen in the next few snippets of code. Let's see how R interprets > first:

```
with(compensation, Fruit > 80)
##  [1] FALSE FALSE FALSE FALSE FALSE FALSE  TRUE
##  [8] FALSE FALSE FALSE FALSE FALSE FALSE FALSE
## [15] FALSE FALSE  TRUE FALSE FALSE FALSE  TRUE
## [22] FALSE FALSE  TRUE FALSE FALSE FALSE  TRUE
## [29]  TRUE  TRUE FALSE FALSE FALSE FALSE  TRUE
## [36] FALSE FALSE FALSE FALSE  TRUE
```

First, **with**() is a handy function ... it says to R, 'LOOK in this data frame, and do what comes next, and then stop looking'. Second, you will notice that the > symbol, a logical, on its own produces a sequence of TRUE and FALSE, identifying *where* in the `Fruit` vector it is TRUE that the value of `Fruit` is > 80. This is handy. Other R functions can use this set of TRUE and FALSE values to retrieve or omit data. This set of TRUE and FALSE values is the information passed to **filter**() ... and this is what **filter**() can act on and return to you.

Using **filter**()

Let's imagine we are interested in all of the trees producing a large amount of fruit. We see from the **summary**() output above that big fruit production means > 80 kg. As with all **dplyr** functions, we first supply the data frame, and then the condition by which we judge whether to return rows (Table 3.1):

```
# find the rows where it is true that Fruit is >80 return
# them as a data frame
filter(compensation, Fruit > 80)

##      Root  Fruit  Grazing
## 1   7.614  87.73 Ungrazed
```

Table 3.1 Some of the more common logical operators and boolean operators, with examples of use in filter().

R logical or boolean	Meaning	Example	Note
"=="	Equals	filter(compensation, Fruit ==80)	The == finds in which rows it is TRUE that the condition is met.
"!="	Does not equal	filter(compensation, Fruit !=80)	The !=finds in which rows it is TRUE that the condition is NOT met.
<, >, >=, <=	Less than, greater than, equal to or greater than, equal to or less than	filter(compensation, Fruit <=80)	
I	OR	filter(compensation, Fruit >80 I Fruit < 20)	OR: in which rows is it TRUE that FRUIT is > 80 OR FRUIT is < 20? It will return all of these rows.
&	AND	filter(compensation, Fruit >80 & Root < 2.3)	AND: BOTH conditions must be true. This will return rows where it is true that both conditions, from two variables, are met.

```
## 2   7.001   80.64 Ungrazed
## 3  10.253  116.05   Grazed
## 4   9.039   84.37   Grazed
## 5   8.988   80.31   Grazed
## 6   8.975   82.35   Grazed
## 7   9.844  105.07   Grazed
## 8   9.351   98.47   Grazed
## 9   8.530   83.03   Grazed
```

We can easily select rows according to multiple conditions. For example, to keep only rows with Fruit > 80 OR less than 20, we employ the boolean *or* symbol |:

```
filter(compensation, Fruit > 80 | Fruit < 20)
```

```
##          Root   Fruit  Grazing
## 1      4.919   14.73 Ungrazed
## 2      5.130   19.28 Ungrazed
## 3      7.614   87.73 Ungrazed
## 4      7.001   80.64 Ungrazed
## 5      4.426   18.89 Ungrazed
## 6     10.253  116.05   Grazed
## 7      9.039   84.37   Grazed
## 8      6.106   14.95   Grazed
## 9      8.988   80.31   Grazed
## 10     8.975   82.35   Grazed
## 11     9.844  105.07   Grazed
## 12     9.351   98.47   Grazed
## 13     8.530   83.03   Grazed
```

3.3.4 MAKING SURE YOU CAN *use* THE SUBSET OF DATA

At the moment, you've been asking R to do stuff, and report the outcome of doing it in the Console. However, very often, you will want to use the results in subsequent jocularities. As you will recall from the previous chapters, the assignment operator (<-) is what you need to use. If we want the low and high Fruit-producing trees for some other activity, we have to assign the result to an object:

```
lo_hi_fruit <- filter(compensation, Fruit > 80 | Fruit < 20)
# now look at it
lo_hi_fruit
```

```
##          Root   Fruit  Grazing
## 1      4.919   14.73 Ungrazed
## 2      5.130   19.28 Ungrazed
## 3      7.614   87.73 Ungrazed
```

```
## 4     7.001   80.64 Ungrazed
## 5     4.426   18.89 Ungrazed
## 6    10.253  116.05   Grazed
## 7     9.039   84.37   Grazed
## 8     6.106   14.95   Grazed
## 9     8.988   80.31   Grazed
## 10    8.975   82.35   Grazed
## 11    9.844  105.07   Grazed
## 12    9.351   98.47   Grazed
## 13    8.530   83.03   Grazed
```

Commit this to memory: *assign the values returned by subsetting using filter to an object (word) if you want to use them again.*

3.3.5 WHAT SHOULD MY SCRIPT LOOK LIKE NOW?

At this point, we'd like to show you what your script should (or might) look like if you were following along. More annotation is always good. Note that annotation can go above, to the right of, or below chunks of code. Note too that we have white space between chunks of code:

```
# my first dplyr script

# clear R's brain
rm(list=ls())

# libraries I need (no need to install...)
library(dplyr)
library(ggplot2)

# get the data
compensation <- read.csv('compensation.csv')

# quick summary
summary(compensation)

# using dplyr; always takes and gives a data frame

# columns
select(compensation, Fruit) # gets the Fruit column
select(compensation, -Root) # take Root column out from data

# rows
slice(compensation, c(2,3,10)) # get 2nd, 3rd & 10th rows

# gets rows for each condition, and assigns to an object
lo_hi_fruit <- filter(compensation, Fruit > 80 | Fruit < 20)

# run this to see what the above line 'saved' for later use.
lo_hi_fruit
```

3.4 Transforming

Transforming columns of your data is a common practice in many biological fields. For example, it is common to log-transform variables for graphing and data analysis. You may also find yourself in a situation where you want to present or analyse a variable that is function of other variables in your data. For example, if you have data on the total number of observations and on the number that were blue, you might want to present the proportion of observations that were blue. Here, we reveal the basic approach to using **mutate**() to achieve these ends.

3.4.1 mutate()

As with all **dplyr** functions . . . **mutate**() starts with the data frame in which the variables reside, and then designates a new column name and the transformation. For example, let's log-transform Fruit, and call it logFruit. We will make this new column appear in our working data frame by employing a neat trick, assigning the values returned by **mutate**() to an object of the *same* name as the original data. We are essentially overwriting the data frame! We will also use **head**() to limit the number of rows we see . . . just for clarity:

```
# what does compensation look like now?
head(compensation)

##      Root Fruit  Grazing
## 1  6.225 59.77 Ungrazed
## 2  6.487 60.98 Ungrazed
## 3  4.919 14.73 Ungrazed
## 4  5.130 19.28 Ungrazed
## 5  5.417 34.25 Ungrazed
## 6  5.359 35.53 Ungrazed

# use mutate
# log(Fruit) is in the column logFruit
# all of which gets put into the object compensation
compensation <- mutate(compensation, logFruit = log(Fruit))

# first 6 rows of the new compensation
head(compensation)

##      Root Fruit  Grazing logFruit
## 1  6.225 59.77 Ungrazed 4.090504
```

```
## 2 6.487 60.98 Ungrazed 4.110546
## 3 4.919 14.73 Ungrazed 2.689886
## 4 5.130 19.28 Ungrazed 2.959068
## 5 5.417 34.25 Ungrazed 3.533687
## 6 5.359 35.53 Ungrazed 3.570377
```

Just to drive home a very nice point about working with R and a script, ask yourself whether you've changed anything in your safe, backed-up, securely stored .csv file on your computer. Have you? Nope. We are working with a copy of our data inside R, manipulating these data, but at no time are we altering the original raw data. You can always go back to those if you need.

3.5 Sorting

3.5.1 arrange()

Sometimes it's important or desirable to put the observations (rows) of our data in a particular order, i.e. to sort them. It may simply be that you'd like to look at the dataset, and prefer a particular ordering of rows. For example, we might want to see the compensation data in order of increasing Fruit production. We can use the **arrange**() function:

```
arrange(compensation, Fruit)
```

```
##     Root Fruit  Grazing logFruit
## 1 4.919 14.73 Ungrazed 2.689886
## 2 6.106 14.95   Grazed 2.704711
## 3 4.426 18.89 Ungrazed 2.938633
## 4 5.130 19.28 Ungrazed 2.959068
## 5 4.975 24.25 Ungrazed 3.188417
## 6 5.451 32.35 Ungrazed 3.476614
```

Another reason for arranging rows in increasing order is if we'd like to perform analyses that need a specific order. For example, some types of time series analyses need the data in the correct temporal order (and may not themselves ensure this). In this case, we would be very well advised to do the sorting ourselves, and check it carefully. (Look at the help file for **arrange**() to see how to sort by multiple variables.)

3.6 Mini-summary and two top tips

So, that was a rapid introduction to five key verbs from *dplyr*. *dplyr* functions are fast and consistent, and do one thing well. On this latter note, here is *Top Tip 1*: you can use more than one *dplyr* function in one line of code! Imagine you want fruit production > 80, and the rootstock widths ONLY. That's a **filter**() and a **select**() agenda, if we've ever heard one:

```
# Root values from Fruit > 80 subset
select(filter(compensation, Fruit > 80), Root)

##      Root
## 1   7.614
## 2   7.001
## 3  10.253
## 4   9.039
## 5   8.988
## 6   8.975
## 7   9.844
## 8   9.351
## 9   8.530
```

Reading this from the inside out helps. Here we've asked for *dplyr* to **filter**() the data first, then take the data frame from **filter**(), and use **select**() to get the Root column only. Nice.

But this leads us to *Top Tip 2*. Built into *dplyr* is a very special kind of magic, provided by Stefan Milton Bache and Hadley Wickham in the *magrittr* package. This gets installed when you install *dplyr*, so you don't need to get it yourself. The magic is found in a symbol called a *pipe*. In R, the pipe command is %>%. You can read this like 'put the answer of the left-hand command into the function on the right'.

Some of you may find this so logical[1] and so much fun that you never stop. We haven't stopped.

Let's translate the two-function code above into 'piped' commands. The art of piping with *dplyr* is to remember to always start with the data frame . . .

---

[1] Some of you may know about the Unix/Linux pipe command '|'. This is modelled on that; the '|' is used in other places in R.

```
# Root values from Fruit > 80 subset
# Via piping
compensation %>%
  filter(Fruit > 80) %>%
    select(Root)

##       Root
## 1   7.614
## 2   7.001
## 3  10.253
## 4   9.039
## 5   8.988
## 6   8.975
## 7   9.844
## 8   9.351
## 9   8.530
```

Read from left to right, top to bottom, this says (1) work with the compensation data, (2) **filter**() it based on the fruit column, getting all rows where it is true that Fruit > 80, and then pass this data frame to (3) **select**() and return *only* the Root column as the final data frame. Sweet.

There are a few reasons we like this. First, it may be nicer and easier to read than putting functions inside functions. Second, this gets even more valuable when more than two functions can potentially be used inside each other. Third, um, well, it is just so cool!

3.7 Calculating summary statistics about groups of your data

At this point, you should be feeling pretty confident. You can import and explore the structure of your data, and you can access and even manipulate various parts of it using five verbs from *dplyr*, including **filter**() and **select**(). What next? In this section, we introduce to you functions that help you generate custom summaries of your data. We will continue working with the compensation data.

In this data frame we have a categorical variable, Grazing. It has two levels, *Grazed* and *Ungrazed*. This structure to the data means we might be able to calculate the *mean* fruit production in each category. Whenever you have structure, or groups, you can generate with R some rather amazing and fast summary information.

The two key *dplyr* functions for this are **group_by()** and **summarise()**. We also introduce the functions **mean()** and **sd()** (standard deviation).

3.7.1 OVERVIEW OF SUMMARIZATION

Summarization is accomplished in a series of steps. The core idea, using *dplyr*, is to:

1. Declare the data frame and what the grouping variable is.
2. Provide some kind of maths function with which to summarize the data (e.g. **mean()** or **sd()**).
3. Provide a nice name for the values returned.
4. Make R use all of this information.

We offer two methods for making this happen.

3.7.2 METHOD 1: NESTED, NO PIPE

In the nested approach, we construct everything as follows. A good test of knowing your data is asking what you might expect. Here, using one grouping variable with two levels and asking for the means, we expect a data frame to be returned with two numbers, a mean for *Grazed* and a mean for *Ungrazed* `Fruit` production:

```
summarise(
  group_by(compensation, Grazing),
          meanFruit = mean(Fruit))

## Source: local data frame [2 x 2]
##
##     Grazing meanFruit
##      (fctr)     (dbl)
## 1    Grazed   67.9405
## 2 Ungrazed   50.8805
```

The second line of code has some good stuff on the 'inside'. The **group_by()** function works with the data frame and declares Grazing as our grouping variable. Of course, if we have more than one grouping variable, we can add them with commas in between. It's that easy.

The third line is where we ask for the mean to be calculated for the Fruit column. We can do this, and R knows where to look, because the **group_by**() function has set it all up. The 'word' meanFruit is some formatting for the output, as you can see in the output.

Beautiful, eh! We get the mean of each of the Grazing treatments, just as expected. And don't forget, if you want to use the means, you must use the <- symbol and assign the result to a new object. Perhaps you'll call it mean.fruit:

```
mean.fruit <- summarise(
  group_by(compensation, Grazing),
        meanFruit = mean(Fruit))
```

3.7.3 METHOD 2: PIPE, NO NESTING

Some of you may have already figured out the piping method. It is perhaps more logical in flow. As above, we always start by declaring the data frame. One big difference is that **summarise**() is now third in the list, rather than on the 'outside'. Start with the data, divide it into groups, and calculate the mean of the fruit data in each group. That's the order:

```
compensation %>%
  group_by(Grazing) %>%
    summarise(meanFruit = mean(Fruit))
```

3.7.4 SUMMARIZING AND EXTENDING SUMMARIZATION

group_by() and **summarise**() are wonderful functions. You can **group_by**() whichever categorical variables you have, and calculate whatever summary statistics you like, including **mean**(), **sd**(), **median**(), and even functions that you make yourself.

In fact, with very few changes to the above code, you can ask for more than one 'statistic' or metric:

```
compensation %>%
  group_by (Grazing) %>%
    summarise(
      meanFruit = mean(Fruit),
      sdFruit = sd(Fruit)
```

This makes creating fast and efficient summaries of your data over many dimensions a doddle.

3.8 What have you learned . . . lots

From Chapter 1 to now, you've gained a great deal of understanding of how R does maths, works with objects, and can make your life easy and *very* organized by allowing you to work in a script. You've learned about tidy data and, importantly, how to import data into R and now know to explore those data. You've gained new skills via ***dplyr***, letting you work with columns, rows, and subsets of your data. Not to mention creating new transformations of your data, and rearranging them. Oh, and that magical pipe.

Furthermore, you've calculated some basic statistics, and in doing so learned several new tricks about working with groups of your data. You are beginning to have the capacity to see the patterns in your data, patterns that you expected. Most importantly, from an R data analysis/data management perspective, you have a *permanent, repeatable, annotated, shareable, cross-platform, executable record* of what you are doing—the script.

Have you saved your script?

We think you are ready for some serious fun now. That serious fun comes from moving from just working with your data, which is super-important, to making figures. Pictures of your data that have meaning, reflect theory, and, if you are lucky, show you the answer to the questions you set out to ask when you were collecting the data. This is what Chapter 4 is all about. See you there.

Appendix 3a   Comparing classic methods and *dplyr*

To be clear, the following is 'possibly nice-to-know' information, and probably not 'need-to-know'. One instance in which you might benefit from it is if you have access to scripts written in your lab long ago, or are working with seasoned R veterans who've not themselves embraced fully the Hadleyverse. A summary comparison of some 'base' methods and *dplyr* methods is given in Table 3.2. A brief description of this comparison follows.

Using *dplyr*, we use the functions **select**(), **slice**(), and **filter**() to get subsets of data frames. The classic method for doing this often involves something called *indexing*, and this is accomplished with square brackets and a comma, with the rows we want before the comma and the columns/variables we want after the comma: something like `mydata[rows,columns]`. There are lots of ways of specifying the rows and columns, including by number, name, or logical operator. It's very flexible, quick, and convenient in many cases. Selecting rows can also be done with the **subset**() base R function, which actually possesses the combined functionality of **filter**() and **select**() from *dplyr*. Ordering rows or columns can be achieved by a combination of indexing and the base R **order**() function.

Adding a new variable, which might be some transformation of existing ones, is also very similar between base R, using the **transform**() function, and *dplyr*, using the **mutate**() function. People often add columns by using a dollar sign followed by the new variable, for example `mydata$new_variable <- mydata$old_variable` (see Table 3.2).

The 'old skool' and still useful methods for getting information about groups of data use functions like **aggregate**() and **tapply**()—both of these were covered in detail in the previous edition of this book. These functions have separate `arguments` that specify the groups and the summary statistics. In *dplyr* the groups are specified by the **group_by**() function and the summary statistics by **summarise**().

Appendix 3b   Advanced *dplyr*

The wonderful depths of *dplyr* are beyond the scope of an introductory R text such as this book. One of our justifications for teaching *dplyr* was,

Table 3.2 Comparison (using the compensation data) of common data manipulation methods using the classic way (base R) and the modern way, with functions in the *dplyr* package. The placement of the comma, when using [ ]'s, is quite important and subtle. The help files for Extract (for [ ]), subset, order, aggregate, and tapply are worth consulting should you run into these functions.

Operation	Base R example	dplyr function	dplyr example
Select some rows	compensation[c(2,3,10),]	slice()	slice(compensation, c(2,3,10))
Select some columns	compensation[,1:2] OR compensation[,c("Root", "Fruit")]	select()	select(compensation, Root, Fruit)
Subset	compensation[compensation $Fruit>80,] OR subset(compensation, Fruit>80)	filter()	filter(compensation, Fruit>80)
Order the rows	compensation[order(compensation $Fruit),]	arrange()	arrange(compensation, Fruit)
Add a column	compensation$logFruit <- log(compensation$Fruit) OR transform(compensation, logFruit=log(Fruit)	mutate()	mutate(compensation, logFruit = log(Fruit)
Define groups of data	-	group_by()	compensation %>% group_by(Grazing)
Summarize the data	aggregate(Fruit ~ Grazing, data = compensation, FUN = mean) OR tapply(compensation$Fruit, list(compensation$Grazing), mean)	summarise() AND group_by()	compensation %>% group_by(Grazing) %>% meanFruit = mean(Fruit)

however, those depths. So we mention a few of them here. And probably not too long after you finish with this book, you'll benefit from knowing a bit about them.

First, an easy one: you can merge together two datasets using one of the join() functions. There are a few versions, because there are a few different ways to merge two datasets, such as keeping all rows, even ones that aren't common between the two datasets, or keeping just the common rows.

Second, and this one is absolutely fantastic, you can apply transformations to groups of your data. You already know how to summarize groups of your data, but this is different. Let's say, for example, you want to subtract the mean of each group of your data from each value in the corresponding group. To do this, you first declare the groups, just as before, using the **group_by**() function. Then you use the **mutate**() function, which you already know as the function for transforming your data. The following code does this, subtracting the mean fruit production for each grazing treatment from the values for the corresponding grazing treatment, using piping, of course:

```
compensation_mean_centred <- compensation %>%
   group_by(Grazing) %>%
      mutate(Fruit_minus_mean = Fruit - mean(Fruit))
```

Magic, eh? Now check that the mean of each group of the Fruit_minus_mean variable is equal to zero.

Third, and finally, you can actually ask for more or less anything to be done to a group of your data. For example, do a linear model for each group. The new function for this is, surprise surprise, **do**(). Here's how we would do a linear model for each grazing treatment:

```
library(broom)
compensation_lms <- compensation %>%
   group_by(Grazing) %>%
      do(tidy(lm(Fruit ~ Root, data=.)))
```

Note that we used a function called **tidy**() from the package ***broom*** to tidy up the output of the linear model function **lm**(). Otherwise, things are rather messy.

There is much, much more one can do with *dplyr*. That is just a taster. Just keep it in the back of your mind. And have a play with one of your own datasets, using that and your scientific question, to drive (constrain) your learning. Otherwise, you may get completely and wonderfully lost in the depths of *dplyr*.

4

Visualizing Your Data

4.1 The first step in every data analysis—making a picture

We advocate a very fundamental rule for data analysis. *Never* start an analysis with statistical analysis. *Always* start an analysis with a picture. Why? If you have done a replicated experiment, conducted a well-organized, stratified sampling program, or generated data using a model, it is highly likely that some expected/hypothesized theoretical relationship underpinned your research effort. You have in your head an *expected* pattern in your data.

Plot your data first. Make the picture that should tell you the answer— make the axes correspond to the theory. If you can do this, *and* see the pattern you *expected*, you are in great shape. You will possibly know the answer!

There are three major figure types that we introduce. The first is the scatterplot, the second is the box-and-whisker plot, and the third is the histogram. Along the way, we are going to show you how to do some pretty special things with colours and lines and *dplyr* too. This will get even more advanced in the chapters on data analysis, and we will come back to figure enhancing in Chapter 8. We note that, in the first edition of this book,

Getting Started with R Second Edition. Andrew Beckerman, Dylan Childs, & Owen Petchey:
Oxford University Press (2017). © Andrew Beckerman, Dylan Childs, & Owen Petchey.
DOI 10.1093/oso/9780198787839.001.0001

we also showed tools to build bar charts ± error bars. However, we since decided that we don't like these, so we changed. Many other people don't like bar charts . . . they can hide too much.[1]

4.2 *ggplot2*: a grammar for graphics

Our plotting in this second edition of *Getting Started with R* is going to centre around **ggplot2**. Not only is **ggplot2** popular, and thus there are immense online resources and help, but it is also extremely effective at working with tidy data and at interfacing with **dplyr**. It is, in a word, awesome. However, it does have a learning curve, so we'll walk you slowly through some basics, and start by graphing a dataset you're already quite familiar with, the compensation.csv dataset.

Let's start with the basic syntax. Here's how we tell the ggplot() function (which resides in **ggplot2**) how to make a simple, plain, bivariate scatterplot of the data:

```
ggplot(compensation, aes(x = Root, y = Fruit)) +
  geom_point()
```

Like the functions in the *dplyr* package, the first argument to give the function ggplot(), the workhorse of plotting, is the data frame (compensation) in which your variables reside. The next argument, aes(), is interesting. First, it is itself a function. Second, it defines the graph's *aesthetics*; that is, the mapping between variables in the dataset and features of the graph. It is here that we tell R, in this example, to associate the *x*-position of the data points with the values in the Root variable, and the *y*-position of the points with the values in the Fruit variable. We are setting up and establishing which variables define the axes of the graph.

The other significant thing to embed in your head about **ggplot2** is that it works by adding layers and components to the aesthetic map. The data and aesthetics *always* form the first layer. Then we add the geometric

[1] http://dx.doi.org/10.1371/journal.pbio.1002128

objects, like points, lines, and other things that report the data. We can also add/adjust other components, like *x*- and *y*-label customization. The trick is to know that the + symbol is how we add these layers and components.

So, in the above example, we follow the first line with a + and then a return/enter. On the next line, we add the geometric layer: points. We use the function **geom_point()**.

This is thus a two-layer graph (Figure 4.1). The first declares the data and aesthetics, and the second pops the points on. If you've actually run this code (having put it into your work-along script), you may or may not like the default output. We'll come back to that. We'll show you a few common and important customizations below. And we provide in Chapter 8 a suite of serious tools and customization magics.

But let's go a bit further with this example . . .

4.2.1 MAKING A PICTURE—SCATTERPLOTS

We suggest at this stage that you prepare a new script. Perhaps call it scatterplot.tutorial.R. Type as many comments (#) as you'd

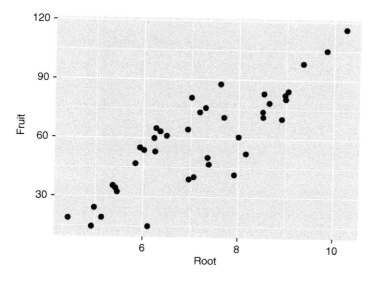

Figure 4.1 A basic scatterplot produced by **ggplot()**.

like, and more than you think you need. Don't forget to start it with **rm**(list=**ls**()). We'll continue to use the fruit and root data—the `compensation.csv` dataset. So get the preliminaries set up:

```
# plotting basics with ggplot
# my tutorial script
# lots and lots of annotation!

# libraries I need (no need to install...)
library(dplyr)
library(ggplot2)

# clear the decks
rm(list = ls())

# get the data
compensation <- read.csv('compensation.csv')

# check out the data
glimpse(compensation)

# make my first ggplot picture
ggplot(compensation, aes(x = Root, y = Fruit)) +
  geom_point()
```

4.2.2 INTERPRETATION, THEN CUSTOMIZE

Before we show you how to do a few basic customizations of this figure, let's have a look at it, biologically. We are interested here in the patterns. Have we recovered what we might expect? Of course. On the *y*-axis are Fruit values and on the *x*-axis the rootstock width. The relationship is clearly positive. *And* there appear to be two groups of points, corresponding to the two groups in the **Grazing** treatment, `Grazed` and `Ungrazed`. Goooood.

Now we are ready to do a few things to make this more representative of the data. We are going to introduce four things: how to quickly get rid of the grey background (everyone wants to know. . .), how to alter the size of points and text, and how to make the colours correspond to the groups for the Grazing factor. As we've noted before, we refer you to Chapter 8 for even more fun and games.

4.2.3 THAT GREY BACKGROUND

This is a divisive feature of **ggplot2**. Don't worry, we don't care if you love it or hate it. Here is a quick way to get rid of the grey. Built into **ggplot2** are a set of themes. One of those is **theme_bw()**. Such themes are good to place as the very last component of your **ggplot()**:

```
ggplot(compensation, aes(x = Root, y = Fruit)) +
  geom_point() +
  theme_bw()
```

There, that was easy. Next, we want to show you a quick way to increase the size of the points. It is as easy as saying that you want the size of the points bigger in the **geom_points()** layer, using a size argument:

```
ggplot(compensation, aes(x = Root, y = Fruit)) +
  geom_point(size = 5) +
  theme_bw()
```

Next, you can certainly alter the *x*- and *y*-axis labels. We modify explicitly here the components **xlab()** and **ylab()**:

```
ggplot(compensation, aes(x = Root, y = Fruit)) +
  geom_point(size = 5) +
  xlab("Root Biomass") +
  ylab("Fruit Production") +
  theme_bw()
```

The final customization we'll show you is how to adjust the colours of the points to match specific levels in a group. This is SO handy for relating figures to statistical models that we just have to show you this. AND it is sooooo very easy with **ggplot2**:

```
ggplot(compensation, aes(x = Root, y = Fruit, colour = Grazing)) +
  geom_point(size = 5) +
  xlab("Root Biomass") +
  ylab("Fruit Production") +
  theme_bw()
```

Can you see the small change? Are you serious? All I need to do, you say, is declare colour = Grazing? Yes. This is a bit of magic. **ggplot()** is mapping the levels of **Grazing** (there are two: Grazed and Ungrazed) to two default colours. You may not like the colours yet, and yes you can change them (Chapter 8), but look how easy it is now to *SEE* the entire structure of the dataset. And, bonus, a graphical legend as well (Figure 4.2).

Oh . . . one more thing, just because it is easy, too. We can *also/or* change the shape of the points to correspond to the Grazing treatment with two levels . . . just by also/or declaring that shape = Grazing:

```
ggplot(compensation, aes(x = Root, y = Fruit, shape = Grazing)) +
  geom_point(size = 5) +
  xlab("Root Biomass") +
  ylab("Fruit Production") +
  theme_bw()
```

So, if you've been doing this with us, you should have a script like this, replete with annotation, please! It is a compact, customized, editable, shareable, repeatable record of your plotting toolset:

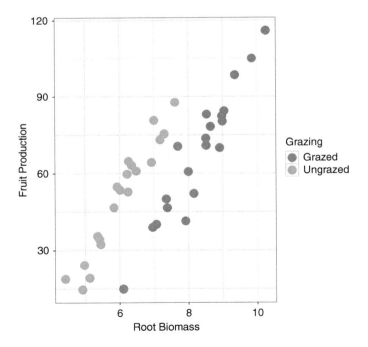

Figure 4.2   A nicely customized scatterplot produced by **ggplot()**.

```
# plotting basics with ggplot
# my tutorial script
# lots and lots of annotation!

# libraries I need (no need to install...)
library(dplyr)
library(ggplot2)

# clear the decks
rm(list = ls())

# get the data
compensation <- read.csv('compensation.csv')

# check out the data
glimpse(compensation)

# make my first ggplot picture
# theme_bw() gets rid of the grey
# size alters the points
# colour and shape are part of the aesthetics
# and assign colours and shapes to levels of a factor
ggplot(compensation, aes(x = Root, y = Fruit, colour = Grazing)) +
  geom_point(size = 5) +
  xlab("Root Biomass") +
  ylab("Fruit Production") +
  theme_bw()
```

4.3 Box-and-whisker plots

Scatterplots, as above, are very good at displaying raw data. However, there are many ideas about presenting data via some version of their central tendency (mean, median) and some estimate of their variation (e.g. standard deviation or standard error). In the biological sciences, bar charts are common. However, recent discussions have suggested that these types of display hide too much information and are not fit for purpose.[2]

In support of this, we are going to suggest using a well-established alternative, the box-and-whisker plot. By the end of this chapter, you'll be able to search online for how to do bar charts if you so desire . . .

[2] See, for example, Beyond Bar and Line Graphs: Time for a New Data Presentation Paradigm, PLOS Biology, http://dx.doi.org/10.1371/journal.pbio.1002128.

Let's look again at the compensation data and focus on how Fruit production (the response variable) varies for each of the Grazing treatments; we'll ignore the Root variable here. **ggplot2** has a built-in geom_ for box-and-whisker plots, not surprisingly called **geom_boxplot()**. Here is how

we can use it:

```
ggplot(compensation, aes(x = Grazing, y = Fruit)) +
  geom_boxplot() +
  xlab("Grazing treatment") +
  ylab("Fruit Production") +
  theme_bw()
```

As with the scatterplot, our first layer declares the data frame and the aesthetics. In this instance, the x-axis aesthetic is a categorical variable, with our two levels. **ggplot()** knows what to do ...

We can also add the raw data as its own layer to this, quite easily. The following highlights a bit more of the extensibility of **ggplot()** and its layering paradigm:

```
ggplot(compensation, aes(x = Grazing, y = Fruit)) +
  geom_boxplot() +
  geom_point(size = 4, colour = 'lightgrey', alpha = 0.5) +
  xlab("Grazing treatment") +
  ylab("Fruit Production") +
  theme_bw()
```

The result of this (Figure 4.3) is a lovely display of the data, median, and spread of the fruit production data. We've snuck a few customization details into the **geom_point()** layer. You can change the size, colour, and *transparency* (alpha) of all the points within this layer, and that's what we have done.

Save your script! This is really important. You do not want to have to type this all in again.

4.3.1 A MOMENT OF INTERPRETIVE CONTEMPLATION ...

We have now produced two spectacular figures. You should be feeling pretty confident. Your *dplyr* and *ggplot2* skills are rocketing. Your foRce is

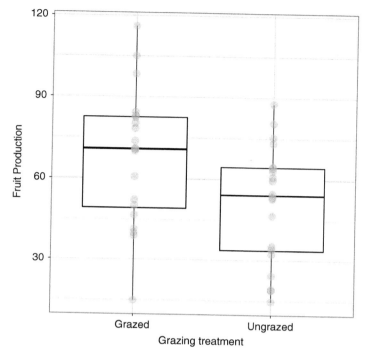

Figure 4.3   A box-and-whisker plot with raw data, produced by **ggplot**().

strong. However, don't forget to look at the graphs very carefully, and bio-logically. Otherwise, why plot them? Have a look at the graphs and answer these questions:

- Do plants with wider root diameters at the start of the experiment produce more fruit?
- Do grazed or ungrazed plants produce more fruit.

4.4 Distributions: making histograms of numeric variables

Looking at the distribution of our variables is extremely important. We can get clues about the shape of the distribution, about the central tendency, about spread, and if there might be some rather extreme values. Many statistical tests can be visualized by looking at the distributions.

Plotting distributions using **ggplot**() involves **geom_histogram**(). It requires you to think a bit about what a histogram is. No, it is not a bar chart. It actually involves some assumptions on the part of the computer, dividing your data into 'bins', then counting observations in each bin, and plotting these counts.

Critically, the above paragraph should indicate that the computer produces the *y*-axis, not you! From a **ggplot**() standpoint, this is important . . . the aesthetics specified in **aes**() only have *one* variable . . . the *x*-variable.

So . . . let's produce a histogram of fruit production:

```
ggplot(compensation, aes(x = Fruit)) +
  geom_histogram()

## `stat_bin()` using `bins = 30`. Pick better value with
## `binwidth`.
```

Now, that (Figure 4.4) doesn't look very nice. The astute observer will notice, having run the code, that **ggplot**() has told you what it did, and how

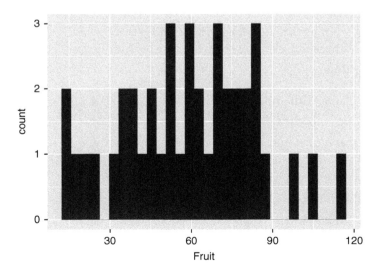

Figure 4.4   An ugly histogram of fruit production.

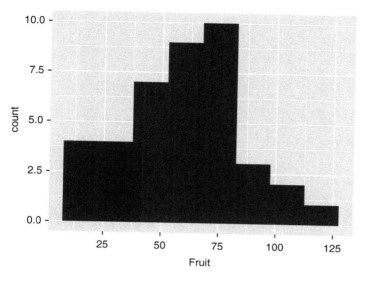

Figure 4.5 A nicer histogram of fruit production.

to fix it—check your Console. We have to change the binwidth. Actually, we can change either the binwidth (how wide each bar is in 'fruit units' (that rolls off the tongue, doesn't it)) or the number of bins (e.g. **ggplot()** defaulted to 30 here). Both of the following produce roughly the same view of the data (Figure 4.5).

```
ggplot(compensation, aes(x = Fruit)) +
  geom_histogram(bins = 10)
ggplot(compensation, aes(x = Fruit)) +
  geom_histogram(binwidth = 15)
```

4.4.1 A NIFTY TOOL: FACETS

We take the opportunity here, with histograms, to introduce one more tool in **ggplot2**. It is a very handy functionality for groups of data. These groups are called facets. R is well known for producing lattice graphics (in fact, the *lattice* package is now part of R's base install). Faceting, or latticing, is about producing a matrix of graphs, automatically structured by a factor/categorical variable in your data.

Keep in mind that this trick works for almost all graphics in **gg-plot2**'s toolbox. But we'll demonstrate it here with the histogram. It

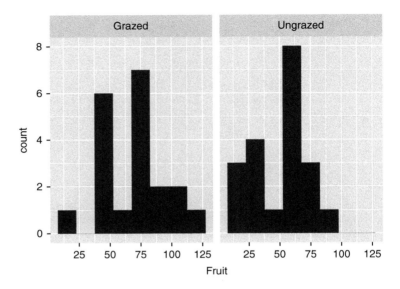

Figure 4.6 Two histograms of fruit production.

requires specifying, yes, a special component of the figure called, wait for it ... **facet_wrap()**. To demonstrate, we take the histogram of fruit data and allow **ggplot()** to divide the data by the **Grazing** treatment, providing *two* histograms for the small price of one bit of code (Figure 4.6). Value.

```
ggplot(compensation, aes(x = Fruit)) +
  geom_histogram(binwidth = 15) +
  facet_wrap(~Grazing)
```

The significant thing to notice about **facet_wrap()** is that the ~ symbol precedes the grouping variable. Locate this symbol. Along with the #, it will become increasingly important.

4.5 Saving your graphs for presentation, documents, etc.

Right. That's three graph types, and some clever colouring and faceting to boot. Clearly, after all this hard work, you want to be able to put the figures in some important places, like presentations, manuscripts, and reports. Let us briefly mention two ways to do this:

1. In the *Plots* tab on the right of RStudio, there is an Export button. This provides options to save to image file types such as .png or .tiff, to save to PDF, or to copy the figure to the clipboard. This works on all platforms via RStudio. A very useful engine of a dialogue box arises, allowing figure size, resolution, and location to be specified.

2. ***ggplot2*** has a built-in classy function called **ggsave()**. **ggsave()** works by saving the figure in the *Plots* window to a filename you specify. Wonderfully, **ggsave()** creates the correct figure type by using the suffix you specified in your filename. For example,

```
ggsave("ThatCoolHistogramOfFruit.png")
```

will save the current figure to the working directory (where R is looking!) as a .png file. Of course, you can define a more complex location, change the height and width, the units, and the resolution, if you so desire. The help file ?ggsave is pretty helpful.

4.6 Closing remarks

Between this chapter and the previous one, you now have in your aRsenal methods for managing and manipulating your data in R, and for making graphs. You can do a great deal with what you just learned, though your dominant current feeling may be that this is all a bit overwhelming. You're right; we just introduced a lot! Probably the best thing you can now do is get a cup of tea or coffee and 7.4 biscuits, and relax. Then, after you've come down off the sugar high, get some of your own data out, get it into R, and have a play with it. Make some graphs. Calculate some summary statistics. Produce a figure related to your data. If you can, see if you can make a picture that answers a particular scientific question. And, at all times, annotate. Use your own words to describe to yourself what you are doing at all stages.

Hopefully you are starting to realize that much of the syntax/vocabulary is quite intuitive (this is supposed to be one of the big advantages of *dplyr* and *ggplot2*). Once you get to grips with the basics, producing gorgeous figures is easy. And, importantly, by capturing your instructions (code) in a script, you can replicate, adjust, and share the figure with anyone, anywhere, and anytime. You can even produce the same figure six months later, within seconds.

What's next? Statistics with R. But don't worry, we'll be making lots of figures with **ggplot**(), and using *dplyr*. Because we always start with a good figure.

Finally, don't forget that we'll cover some more advanced customizations of graphs and more advanced features of the *ggplot2* in Chapter 8.

5

Introducing Statistics in R

5.1 Getting started doing statistics in R

We would like to emphasize what we started with in Chapter 4: we have
a very fundamental rule for data analysis. *Never* start an analysis with
statistical tests. *Always* start an analysis with a picture.

Why? As we noted, you should have in your head an *Expected* pattern in
your data and a picture is a good way to see whether your data meet these
expectations. Plot it first. Make the picture that should tell you the answer.
Now, what do you do after you plot the data? Our philosophy, translated
into a workflow, is as follows.

Once you've made a picture, you embark on translating the hypothesis
you are testing into a statistical model and this model into R language. If
you've made an informative picture that reveals the relationships among
variables central to your hypothesis, this should be easy (or become easier
with experience).

Once you've specified your model in R and made R build it, we feel
that it is vital *not* to then interpret the results. It is, instead, vital to assess first the assumptions of your model. For example, a two-sample *t*-test
may assume equal variance in the two groups, and ANOVA and regression
assume normally distributed residuals, amongst other things. By assessing

Getting Started with R Second Edition. Andrew Beckerman, Dylan Childs, & Owen Petchey:
Oxford University Press (2017). © Andrew Beckerman, Dylan Childs, & Owen Petchey.
DOI 10.1093/oso/9780198787839.001.0001

if these assumptions are met, you are ensuring that the results returned by the modelling are reliable. If the assumptions are not met, then the predictions from, or interpretation of, the model is compromised. You will not be making reliable inference.

Only once you have assessed the assumptions should you begin to interpret the output of the statistical modelling. It is here that we interpret the test statistic and associated *p-value*. The final step is to integrate the modelling results into your original figure—a process some of you may know as adding predicted or fitted lines (or points) to your graph.

This chapter is a long one and covers four simple statistical tests in discrete chunks. Take breaks as you work through the sections. In each, we follow this workflow: (1) plot your data, (2) build your model, (3) check your assumptions, (4) interpret your model, and (5) replot the data *and* the model. Along the way, we aim to reinforce several principles from Chapters 1–4, taking advantage of the tools and skills you developed with **dplyr** and **ggplot2**. Our aim with these examples is to show you how easy interpreting statistics can be if you can make an informative picture before you embark on the analysis. Just to whet your appetite, Chapters 6 and 7 expand the types of tests, introducing the two-way ANOVA and ANCOVA in Chapter 6 and then the generalized linear model in Chapter 7.

5.1.1 GETTING READY FOR SOME STATISTICS

If you have never actually used a *t-test*, χ^2 *contingency table analysis, linear regression* or *one-way ANOVA*, or never taken a course in statistics that introduced these tools, now is a good time to go and acquire a bit of knowledge. Our instructions will certainly teach you some statistics, but our goal is to teach you how to use (get) R to do these. We are *assuming* that you understand when and for what type of data you may require these tools. We are *assuming* that you understand what types of *hypotheses* can be tested with these different methods.

5.2 χ^2 contingency table analysis

We wish to make two points with this introduction to the χ^2 ('chi-squared') contingency table analysis. First, always plot your data first. Did you expect that? Second, make every attempt to understand the hypothesis that you are testing, both biologically and statistically.

The χ^2 contingency table analysis is an analysis of count data. It is essentially a test of association among two or more categorical variables. The need for this sort of analysis arises when you have a set of observations, or events, and for each of these you can *classify* them by more than one categorical variable with, for example, two levels (for example, sex, male/female, and life stage, juvenile/mature). Introducing some data will help explain the basics.

5.2.1 THE DATA: LADYBIRDS

Let's assume that we have collected data on the frequency of black and red ladybirds (*Adalia bipunctata*) in industrial and rural habitats. Can you already see the two grouping variables, each with two levels?

Why might we have collected such data? It has long been thought that industrial and rural habitats, by virtue of the amount of pollution in them, provide different backgrounds against which many insects sit. Contrasting backgrounds to the insect can be bad if contrast is associated with predation risk. Here, we are interested in whether dark morphs are more likely to reside on dark (industrial) backgrounds. These data are available at http://www.r4all.org/the-book/datasets; if you have already downloaded the zipped-up file of all datasets, you already have it. The data file is ladybirds_morph_colour.csv.

By performing a χ^2 contingency table analysis, we are testing the null hypothesis that there is no association between ladybird colours and their habitat. The alternative is that there is. The test does not allow us to specify the direction of this association, but that is something we will see from the graph we make before performing the test. Biologically, we are trying to answer the question of whether some feature of the habitat is associated with the frequencies of the different colour morphs. Note how generic this statement is—there is no specification of the features of the habitat, or of the direction of potential association.

Now we know what questions we're asking, let's read in the data and check their structure. Remember to set up a new script, clear R's brain, define the packages you want to use, set the working directory, and get the data:

```
# My First Chi-square contingency analysis

# Clear the decks
rm(list = ls())

# libraries I always use.
library(dplyr)
library(ggplot2)

# import the data
lady <- read.csv("ladybirds_morph_colour.csv")

# Check it out
glimpse(lady)


## Observations: 20
## Variables: 4
## $ Habitat       (fctr) Rural, Rural, Rural, Rural, Rural...
## $ Site          (fctr) R1, R2, R3, R4, R5, R1, R2, R3, R...
## $ morph_colour  (fctr) black, black, black, black, black...
## $ number        (int) 10, 3, 4, 7, 6, 15, 18, 9, 12, 16,...
```

Looking at the output of **glimpse**(lady), we see that the dataset is quite *tidy*, with 20 rows, each of which is a location at which ladybirds were observed. The Habitat variable shows whether the observation was in an *Industrial* or *Rural* location, the Site variable is a unique identifier for the location, the morph_colour variable is which morph colour was observed (red or black), and the number variable is the number of individual ladybirds of that morph colour that were observed.

(Note that it is a good idea to be consistent with the case of variable names, rather than having some with an upper-case first letter and others with a lower-case first letter. We like to keep you on your toes, though.)

There are various ways we could analyse these data to answer our question. We're going to take a quite simple approach (for learning purposes): a χ^2 contingency table analysis, to test if the frequencies of the two colours of ladybirds differ between the two habitat types.

The first thing you may notice is that we first need to calculate these totals ... This is a chance to make use of your *dplyr*-ninja skills. Before you do this, think about what we are aiming for ... four numbers, each of which is the sum of the observations of each colour in each habitat. Make sure you understand that.

5.2.2 ORGANIZING THE DATA FOR PLOTTING AND ANALYSIS

If you understand our aim in summarizing the data, you probably realize that you can use the **group_by()** and **summarise()** functions in the *dplyr* package:

```
totals <- lady %>%
  group_by(Habitat, morph_colour) %>%
    summarise(total.number = sum(number))
```

Making a figure of these data is easy. And this is the *only* time we advocate using a bar chart to summarize your raw data (see below). Here's how to do it:

```
ggplot(totals, aes(x = Habitat, y = total.number,
       fill = morph_colour)) +
  geom_bar(stat = 'identity', position = 'dodge')
```

5.2.3 NEW **ggplot()** DETAIL

This figure we've made (Figure 5.1) needs some explanation, doesn't it? How can it be that we just told you never to make a bar chart, and here we are making a bar chart? In the case of count data like this, each bar is one

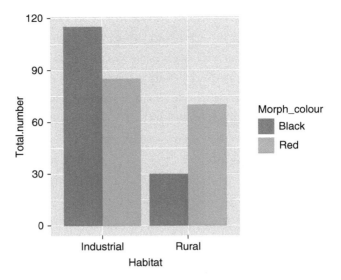

Figure 5.1 Total numbers of red and black ladybirds collected in industrial and rural habitats. Default R colours.

point, we're not summarizing a distribution, and the counts are ratio data (they have a 'real' zero). It *is* a logical way to view these data, particularly as they have (have you noticed?) no 'variation' to examine. In this situation, we think a bar chart starting at zero is actually a good option.

The **ggplot**() syntax has some new things to which to pay attention. First, we see a new part of the aesthetic, fill = morph_colour. This is used when the 'geometry' is something like a bar, and can be filled. Incidentally, colour = morph_colour with a bar alters the *outline* of the bar, not the fill colour. Remember that if you want to avoid future frustration.

Second, **geom_bar**() has some arguments that are worthy of explanation. stat = 'identity' tells ggplot *not* to try and calculate anything from the data. **ggplot**() *can* do stuff if you let it, but in this case we want **ggplot**() to use *just what we've given it to use*. position = 'dodge' is not a reference to any games or US cities. It is a request to put the two bars in each Habitat group (e.g. black and red counts) next to each other. If you don't use the position = 'dodge' option you'll end up with a stacked barplot.

5.2.4 FIXING THE COLOURS

The colours in Figure 5.1 were chosen by R, and are far from ideal. Black is red, and red is blue!!! Let's make black colour correspond to black lady-birds, and red to red ones. That is, we customize the colours used to represent groups of data.

When working with discrete groups (as in the variables Habitat and colour_morph), *ggplot2* has built-in scale_ functions ending with _manual. We'll provide way more info about scale_'s in Chapter 8. Here we use scale_fill_manual; it takes an argument called values that is a set of colours to use:

```
ggplot(totals, aes(x = Habitat, y = total.number,
     fill = morph_colour)) +
  geom_bar(stat = 'identity', position = 'dodge') +
  scale_fill_manual(values = c(black = "black", red = "red"))
```

Lovely (Figure 5.2)! Note that, in this case, we specify values = c(black = "black", red = "red"). This seems rather obvious, perhaps, but we are

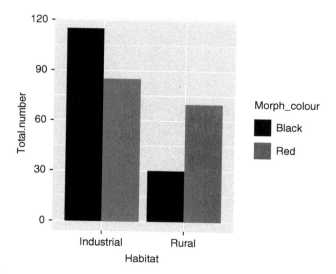

Figure 5.2 Total numbers of red and black ladybirds collected in industrial and rural habitats. Colours manually specified.

assiging the colours to the appropriate morph_colour. If, for example, the morphs of the ladybirds were large and small, and we wanted to use 'black' for large and 'red' for small morphs, we would use values = c(large = "black", small = "red").

5.2.5 INTERPRETING THE GRAPH (GUESS THE ANSWER BEFORE WE DO STATS)

Right... now back to biology. At this point, we can ask ourselves whether we think the null hypothesis—no association between colour morphs and habitat, the same relative abundance of black and red morphs in each habitat—is true or not. Look at Figure 5.2, and decide for yourself.

It certainly looks like the black morph is more common in the industrial habitat, relative to the red morph. Can you see this? Make sure you can—if you are confused, remember that a χ^2 test compares the frequencies/proportions, not the absolute numbers. This suggests that the morphs are *not* equally distributed among the habitats. We are expecting to *reject* the null hypothesis.

5.2.6 MAKING THE χ^2 TEST

To actually test this hypothesis, we need to do two things. Step 2 is to use the function **chisq.test**(). But, to make this function do a two-way contingency test, we must give it a matrix of the total counts. Although they can look superficially similar, a matrix is different from a data frame. At present, because all *dplyr* functions take and give back a data frame, we have all four totals in a single column of a three-column data frame:

```
totals

## Source: local data frame [4 x 3]
## Groups: Habitat [?]
##
##      Habitat morph_colour total.number
```

```
##           (fctr)         (fctr)         (int)
## 1 Industrial         black           115
## 2 Industrial           red            85
## 3      Rural         black            30
## 4      Rural           red            70
```

We introduce one trick here that transforms this data frame into the matrix we need for the χ^2 test: the function **xtabs()**. If you have used Excel, you may be familiar with a pivot table for doing cross-tabulation. **xtabs()** is cross(x)-tabulation:

```
lady.mat <- xtabs(number ~ Habitat + morph_colour,
                  data = lady)
lady.mat

##              morph_colour
## Habitat        black red
##    Industrial    115  85
##    Rural          30  70
```

xtabs() requires a simple formula and a data frame. The data frame in this situation is the raw data from above, lady (although it would also work just fine with the summarized data, totals). The formula reads, '*Please cross-tabulate* the number column of counts in the totals data frame by the Habitat and morph_colour variables.' This carries out exactly the same calculation as the **group_by()** and **summarise()** functions did to make totals, but now we end up with a matrix. Nice.

Now the χ^2 test:

```
chisq.test(lady.mat)

##
##   Pearson's Chi-squared test with Yates' continuity
##   correction
##
## data:  lady.mat
## X-squared = 19.103, df = 1, p-value = 1.239e-05
```

This rather sparse output provides all the information we need in order to present our 'result'. According to the test, there is a very small probability (p = 0.00001239) that the pattern we see is consistent with the null hypothesis; that is, if there really were no association between colour morph and habitat, and we carried out the sampling process again and again, we would get such a similar (or more extreme) result only once every 10,000 or so samples. This is a pretty good indication that it probably isn't a chance result. The result allows us to reject the null hypothesis and conclude that there is some association. Our figure suggests that black morphs are more frequent in industrial habitats, while red morphs are more frequent in rural habitats.

What might we report in a manuscript? 'We tested the hypothesis that there is an association between colour morphs of ladybirds and industrial and rural habitats. Ladybird colour morphs are not equally distributed in the two habitats (x^2 = 19.1, df = 1, p < 0.001), with black morphs being more frequent in industrial habitats and red morphs more frequent in rural habitats (Figure 5.1).'

By the way, if you are wondering what that 'Yates' continuity correction' is all about, it refers to a little adjustment of the x^2 test that makes it a bit more reliable when the counts are small. For those of you who are familiar with the mechanics of the x^2 test (row sums, column sums, observed and expected values), you will be pleased to know that all of these are accessible to you, simply by assigning the values returned by **chisq.test**() to a name and then looking at these:

```
lady.chi <- chisq.test(lady.mat)
names(lady.chi)

## [1] "statistic" "parameter" "p.value"   "method"
## [5] "data.name" "observed"  "expected"  "residuals"
## [9] "stdres"

lady.chi$expected

##               morph_colour
## Habitat           black        red
##    Industrial  96.66667 103.33333
##    Rural       48.33333  51.66667
```

5.2.7 FROM DATA TO STATISTICS: AN OVERVIEW

We had two main points to make with this introduction to the χ^2 contingency table analysis: to remind you to plot your data first and to implore you to think about how to translate the biological question you are asking into a statistical hypothesis. Hopefully you've grasped the importance of each, and we have reinforced some basic details about the χ^2 contingency table analysis. Most importantly, this example should have shown you the ease with which you can gain understanding of your data using R. You now have a script/recipe for processing, graphing, and analysing data such as these ... you have saved your script, yes?

5.3 Two-sample *t*-test

As above, we wish to make three points with this introduction to the two-sample *t*-test. First, always plot your data. Second, check model assumptions—vital for reliable interpretation. Finally, R makes all of this easy.

The two-sample *t*-test is a comparison of the means of two groups of numeric values. It is appropriate when the sample sizes in each group are small. However, it does make some assumptions about the data being analysed. The standard two-sample, Students *t*-test assumes that the data in each group are normally distributed and that their variances are equal. We will come back to these issues shortly but, immediately, you might think about the graph you want to make of these data, showing the distributions. Perhaps two histograms?

5.3.1 THE *t*-TEST DATA

First, some data. Here we will analyse ozone levels in gardens distributed around a city. The gardens were either to the west of the city centre or to the east. The data are ozone concentrations in parts per hundred million (pphm). Ozone in excess of 8 pphm can damage lettuce plants—they bolt and get filled with latex and are yucky to eat. We are interested in whether there is a difference in the average ozone concentration between gardens in the east and the west.

As above, let's read in the data and check their structure. The `ozone.csv` data can be found in the collection of datasets at `http://www.r4all.org/the-book/datasets` —you've probably already downloaded it. Remember to set up a new script, clear R's brain, define the packages you want to use, set the working directory, and get the data.

Following the convention we established above, we assign the data to a name (object), called ozone. Once the data are into R (using read.csv), examine them using one of our suggested tools, such as **glimpse**(). Refer to the example script above to organize the first few lines:

```
glimpse(ozone)

## Observations: 20
## Variables: 3
## $ Ozone           (dbl) 61.7, 64.0, 72.4, 56.8, 52.4, 4...
## $ Garden.location (fctr) West, West, West, West, West, ...
## $ Garden.ID       (fctr) G1, G2, G3, G4, G5, G6, G7, G8...
```

The output from **glimpse**(ozone) reveals that our data are in a data frame format with three columns and that Garden.location is a factor with two levels. R has assumed (because they are not numbers) that the data in the **Garden.location** column should be treated as a factor—a code denoting a categorical grouping variable within the dataset.

5.3.2 THE FIRST STEP: PLOT YOUR DATA

The first step in an analysis of data is ... making a figure! We can see that there are two different locations of gardens: East and West. To initiate a comparison of ozone levels between East and West gardens, we might consider a plotting method that allows us to see the central tendency of the data and the variability in the data, for each location. Since there are only two groups, a good tool for this is the histogram.

Recall that in Chapter 4 we introduced the histogram *and* facets, a tool to break data into groups during the plotting process. Let's take advantage of this. Feel free to first check back in Chapter 4.

If we stack the histograms on top of each other, we should be able to (a) see whether the means seem different and (b) assess whether the data in each location appear to be normal and have similar variance. We thus accomplish the process of visualizing our hypothesis and evaluating some assumptions of a two-sample t-test, all in one effort:

```
ggplot(ozone, aes(x = Ozone)) +
  geom_histogram(binwidth = 10) +
  facet_wrap(~ Garden.location, ncol = 1) +
  theme_bw()
```

The resulting Figure 5.3 provides visual representations of the distributions of the data in each sample, and from these it seems reasonable to say that the assumption of normality and equality of variance has been met. Let us assume that it has. Of course, R has functions to statistically evaluate normality and equality of variance. And we bet you can guess what ncol = 1 means...

The second reason for plotting the data like this was to provide some indication of whether our null hypothesis—that there is no difference in the mean ozone levels between the two groups—is true or not. A cursory

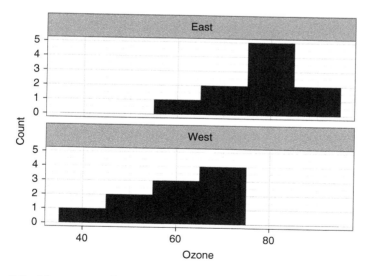

Figure 5.3 Histograms of ozone levels in gardens in East and West garden locations.

glance at the figure we've produced might suggest that maybe we will be able to reject the null hypothesis, i.e. the peaks of the two histograms are at different locations on the *x*-axis. However, there is also considerable overlap between the two histograms. We need some stats. Perhaps at this point you might generate some code, using **dplyr** (e.g. **group_by**() and **summarise**()) to calculate the means and standard errors of the ozone levels in each location? The previous chapters have equipped you to do this.

5.3.3 THE TWO-SAMPLE *t*-TEST ANALYSIS

 We now have an informative figure and have made an informal assessment of whether we might reject the null hypothesis. If you've not done this, go do it! To do the *t*-test, use the **t.test**() function in R. You may even want to read the help file! Add the following to your script:

```
# Do a t.test now....
t.test(Ozone ~ Garden.location, data = ozone)
```

First, let's look at the arguments we give to the **t.test**() function: we provide a 'formula' and a data argument. The data argument specifies the data frame in which the variables reside. We know that. The formula is a representation of our hypothesis: do ozone levels (**Ozone**) vary as a function of location (**Garden.location**)? That is how the ~ makes us read the formula.

Let us spend a bit of time reviewing the output as a way to reinforce aspects of our workflow. We will call this the anatomy of the R output:

```
##
##   Welch Two Sample t-test
##
## data:  Ozone by Garden.location
## t = 4.2363, df = 17.656, p-value = 0.0005159
## alternative hypothesis: true difference in means is not equal to 0
## 95 percent confidence interval:
##    8.094171 24.065829
## sample estimates:
## mean in group East mean in group West
##                77.34                61.26
```

The first line of the output tells us the statistical test performed: a Welch two-sample *t*-test. This is what we expected, apart from the word 'Welch': this is a clear indication that R is doing something slightly unexpected, and we should not ignore such an alarm bell. We'll cover this later, however.

Next we see that the data that have been used are declared—this is a good way to confirm that you've analysed the data you wanted to analyse. The next line provides the traditional *t*-test statistic, degrees of freedom, and *p*-value. We assume you know what these mean. However, let us use the next few lines of output to make a few clear points about the two-sample *t*-test. The output declares the alternative hypothesis to us: that the true difference in means is not equal to 0. This should help you understand more of what you've been doing. If the difference between two things is 0 then they are the same, and we have no grounds for rejecting the null hypothesis.

Next in the output is a 95% confidence interval. This interval is around *the difference between the two means.* Keep in mind that the difference would be 0 if the means were the same. The fact that this interval does not include 0 provides an answer to our initial question. We can conclude that they are probably different. This falls in line with the test statistic and associated *p*-value. Finally, the output provides the means in each group.

Now, back to the word 'Welch'. A quick look in the help files (?t.test) or on Wikipedia reveals that this method allows one of the assumptions of the standard two-sample *t*-test to be relaxed—that of equal variance. While you may have made the effort to assess the assumption of equal variance above, and found the variances relatively similiar, you now know that there are options for when this assumption is not met!

You might have once been taught to test for equality of variance in a two-sample *t*-test. It isn't necessary to do this when using the Welch version—and, actually, we don't think it is ever a very good idea—but if feel you must make a formal test, there are several functions you could use.

For example, the **var.test()** function has the same structure as the *t*-test function:

```
var.test(Ozone ~ Garden.location, data = ozone)
```

5.3.4 *t*-TEST SUMMARY

A two-sample *t*-test is often assumed to be easy. If you can collect data in a manner that allows you to test a fundamental hypothesis using *t*-tests, this is certainly a good thing. But don't let the workflow of a good analysis slip just because it is simply a comparison of two groups: always plot your data, evaluate assumptions, and only then interpret your analysis results.

5.4 Introducing ... linear models

The previous two sections focused on relatively simple questions, and we continue doing that here, but moving into models known as 'general linear models'. General linear models are a class of model that includes regression, multiple regression, ANOVA, and ANCOVA. All of these are fundamentally linear models. These models share a common framework for estimation (least squares) and a common set of assumptions, centred around the idea of normally distributed residuals. The next two examples introduce the assessment of these assumptions as part of our *Plot -> Model -> Check Assumptions -> Interpret -> Plot Again* workflow.

A moment to anticipate confusion. Don't confuse the *general* linear model with the *generalized* linear model, known as the GLM. We introduce the GLM, where key assumptions about normality are relaxed, in Chapter 7.

Perhaps the best way forward is just to get stuck in to the next two examples. We will be using several new functions, but the most important one is **lm()**, which, by the looks of it, is a tool to fit linear models.

5.5 Simple linear regression

The example we are going to use is one that asks whether plant growth rates vary with soil moisture content. The underlying prediction is that more moisture will likely allow higher growth rates (we like simple examples). Two features of these data are important. The first is that we have a clear relationship specified between the two variables, one that is easily visualized by plotting the response (dependent) variable—plant growth rate—against the explanatory (independent) variable—soil moisture content. The second is that the explanatory variable is a continuous, numeric variable. It doesn't have categories.

5.5.1 GETTING AND PLOTTING THE DATA

The dataset is available from the same place as all the previous ones; you already have it. It is called `plant.growth.rate.csv`. Go ahead and get these data, start a linear models script, clear the decks, make the **dplyr** and **ggplot2** packages available, and import the data.

If all goes to plan, you should be able to see the data via **glimpse()** or any of the other tools we introduced in Chapters 2 and 3. We'll call the data frame plant_gr:

```
glimpse(plant_gr)

## Observations: 50
## Variables: 2
## $ soil.moisture.content (dbl) 0.4696876, 0.5413106, 1.6...
## $ plant.growth.rate     (dbl) 21.31695, 27.03072, 38.98...
```

We can see that the structure of the data is as expected, with two continuous variables. Making a figure of these data involves producing a scatterplot, using **geom_point()**. Not much to it, really. Feel free to change colours or point sizes, just to practice some of the basic **ggplot2** tools we've

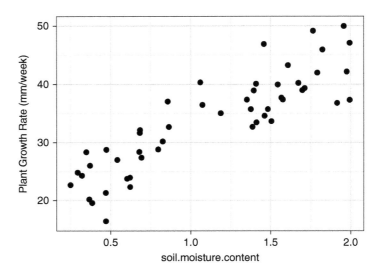

Figure 5.4   An exploratory graph needn't be beautiful. Clarity and speed are priorities.

introduced. We've added one annotation—introducing units of mm/week to the growth rate data (Figure 5.4):

```
ggplot(plant_gr,
       aes(x = soil.moisture.content, y = plant.growth.rate)) +
       geom_point() +
       ylab("Plant Growth Rate (mm/week)") +
       theme_bw()
```

5.5.2 INTERPRETING THE FIGURE: BIOLOGICAL INSIGHT

Looking at the resulting Figure 5.4, there are a few things that you, a budding data scientist and interpreter of all things data, can do. First, you should realize that the gradient (i.e. the slope) is positive. Indeed, the more moisture in the soil, the higher the plant growth rate. Good. Biology seems to be working.

Second, you may not realize this, but you can actually pre-empt the statistical analysis by guestimating the slope and intercept. Check it out. *Roughly speaking*, the growth rate varies between 20 and 50 mm/week.

Roughly speaking, the soil moisture varies between 0 and 2. *Thus*, the gradient is, *roughly speaking*, 30/2 = 15. And the intercept is somewhere, *roughly speaking*, between 15 and 20 mm/week. Gooood. This is always a good idea. Try and examine your data *before* you do the analysis. Push yourself even further by figuring out, in advance, the degrees of freedom for error that you expect. Hint: it's the number of data points minus the number of parameters (also called coefficients) estimated.

5.5.3 MAKING A SIMPLE LINEAR REGRESSION HAPPEN

This next step is rather straightforward. We use the function **lm()** to fit the model. The **lm()** function is very much like xtabs and t.test; it needs a formula and some data:

```
model_pgr <- lm(plant.growth.rate ~ soil.moisture.content,
                data = plant_gr)
```

This reads, 'Fit a linear model, where we hypothesize that plant growth rate is a function of soil moisture content, using the variables from the plant_gr data frame.' Nice. And. Simple.

5.5.4 ASSUMPTIONS FIRST

Now, we know you want to rush ahead and find out whether our a priori estimates of the intercept and gradient/slope were correct. And whether, as aeons of biology would suggest, growth rate increases with soil moisture content. But WAIT, we say. Do not rush excellence.

First, check the assumptions of the linear model. ***ggplot2*** needs a little help with this, as it doesn't know what linear models are. Help comes from ***ggfortify***, and its **autoplot()** function, which, when given a linear model created by **lm()**, produces four very useful figures (Figure 5.5). For those of you familiar with base graphics or with the previous version of the book, these are the same four pictures that **plot()** produces when it is provided with an **lm()** model. We suggest that after installing ***ggfortify***, you add library(ggfortify) to your list of libraries at the start of every script.

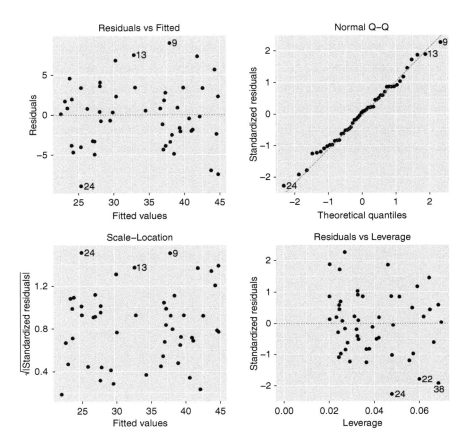

Figure 5.5   Nothing looks too bad in the diagnostic plots.

Here's what you need to do to produce the plots now:

```
library(ggfortify)
autoplot(model_pgr, smooth.colour = NA)
```

What, pray tell, do these plots mean? We actually hope you know what some of them are. They are all based around the residuals—errors around the fitted line. We assume you *do* know what residuals are, and perhaps their relationship to sums of squares and mean square errors. We'll revisit the smooth . colour = NA argument below.

Here we go:

1. *Top left*. This panel is about the 'systematic part' of the model; it tells us whether a line is appropriate to fit to the data. If things have gone wrong, hump-shapes or valleys will be apparent. These would mean that the structure of your model was wrong. For example, fitting a straight line didn't work. Lots of people suggest you look at this to evaluate the assumption of equal variance—homo- vs heteroskedasticity. But there is a better (bottom left) for this.

2. *Top right*. This evaluates the assumption of normality of the residuals. The dots are the residuals, and the dashed line the expectation under the normal distribution. This is a *much* better tool than making a histogram of the residuals, especially with small sample sizes ... like less than 100.

3. *Bottom left*. This evaluates the assumption of equal variance. The *y*-axis is a standardized (all positive) indicator of the variation. Linear models assume that the variance is constant over all predicted values of the response variable. There should be no pattern. But there might be one if, for example, the variance increases with the mean, as it might with count data (see Chapter 7).

4. *Bottom right*. This evaluates leverage, a tool not only to detect influential data points, ones that move the gradient more than might be expected, but also to detect outliers. Your friends, mentors, or supervisor might think this is important. If they do, speak with them ...

If you still feel confused, don't know what these mean, or have relatively little experience with assessing the assumptions the of a general linear model, head for the stats books such as those of Crawley (2005 and 2012), Faraway (2014), and Dalgaard (2008) (see Appendix 2, 'Further Reading', for details of these books).

Overall, the take-home message from this example, and these plots, is that everything is just fine. There is no pattern in either of the left-hand plots. The nomal-distribution assumption is clearly met. And there are no points exerting high influence. Of course, you can test for normality (e.g. by the Kruskal–Wallis or Anderson–Darling test), but we leave that to you, should you wish to or be told to pursue these tests (you might have guessed by now that we're not a fan of them).

As promised, we will tell you about smooth.colour = NA. In the absence of this argument, the default presentation of the diagnostic plots includes a 'wiggly line' fitted by locally weighted regression. The = NA suppresses the line. Many people find this line useful, but we have found that they also tend not to look at the data points. Furthermore, the lines apparently talk to them and tell them there are problems when there really are none. We find, in a nutshell, that omitting these lines is quite beneficial. When there are problems, you can see them without the (un)helpful line (see Chapter 7).

5.5.5 NOW THE INTERPRETATION

Phew. Now we are ready. Ready to see whether we can reject the null hypothesis that soil moisture has no effect on plant growth rate. Ready to see if our guesses were right, to assess the real estimate of plant growth at zero soil moisture (the intercept) and the change of growth rate with soil moisture (the gradient/slope).

We do all of this using two tools, tools we will use for every general (and generalized) linear model from here on in and out: **anova**() and **summary**().

Now, wait a minute, you say. **anova**()? Yes. **anova**(). Repeat after us: **anova**() does not perform an ANOVA. Or, more accurately, it does not carry out the type of ANOVA that compares means. Again? OK. You get it. **anova**() produces a classic table in statistics, the sums-of-squares table. It provides the overall F-value for the model, representing the ratio of variance explained by the explanatory variables to the leftover variance. It also produces an estimate of R^2 and the adjusted R^2.

summary() is far less contentious. It produces a table of the estimates of the coefficients of the line that is 'the model': an intercept and a slope. Of course, there is a bunch of other stats stuff along with that too. But let's see how they all come together.

First the **anova**() table:

```
anova(model_pgr)

## Analysis of Variance Table
##
## Response: plant.growth.rate
##                          Df  Sum Sq Mean Sq F value    Pr(>F)
## soil.moisture.content     1 2521.15 2521.15  156.08 < 2.2e-16
## Residuals                48  775.35   16.15
##
## soil.moisture.content ***
## Residuals
## ---
## Signif. codes:
## 0 '***' 0.001 '**' 0.01 '*' 0.05 '.' 0.1 ' ' 1
```

If you've had any course in stastistics (we hope you have), this should be familiar. The ANOVA table is a classic assessment of the hypothesis, presenting the F-value, degrees of freedom, and p-value associated with the explanatory variables in the model (in this case, the model has one explanatory variable).

We can see here a rather large F-value, indicating that the error variance is small relative to the variance attributed to the explanatory variable. This, with the single degree of freedom, leads to the tiny p-value. As we noted in the χ^2 example, if there really was *no* relationship between plant growth rate and soil moisture, and we were to carry out the sampling process again and again, we would get such a large F-value fewer than one in a million or more samples. This is a pretty good indication that the pattern we are seeing probably isn't a chance result.

And now the **summary**() table:

```
summary(model_pgr)

##
## Call:
```

```
## lm(formula = plant.growth.rate ~ soil.moisture.content,
      data = plant_gr)
##
## Residuals:
##     Min      1Q  Median      3Q     Max
## -8.9089 -3.0747  0.2261  2.6567  8.9406
##
## Coefficients:
##                        Estimate Std. Error t value Pr(>|t|)
## (Intercept)              19.348      1.283   15.08   <2e-16
## soil.moisture.content    12.750      1.021   12.49   <2e-16
##
## (Intercept)            ***
## soil.moisture.content  ***
## ---
## Signif. codes:
## 0 '***' 0.001 '**' 0.01 '*' 0.05 '.' 0.1 ' ' 1
##
## Residual standard error: 4.019 on 48 degrees of freedom
## Multiple R-squared:  0.7648, Adjusted R-squared:  0.7599
## F-statistic: 156.1 on 1 and 48 DF,  p-value: < 2.2e-16
```

The estimates provided in the summary table (the first column) correspond to the estimates, in a linear regression, of the intercept and slope
associated with the explanatory variable. Can you guess which is the intercept... More seriously, make sure you understand that the gradient/slope
is associated with the explanatory variable, in this case soil moisture, which
is also the x-axis of our figure, the values of which are associated with
differences in plant growth rate.

Recalling our guesses from the figure we made, we've done pretty well.
Roughly speaking, we predicted a gradient/slope of 30/2 = 15 vs 12.7 estimated by least squares, and an intercept between 15 and 20 mm/week vs
19.34 estimated by least squares. Making that figure makes our statistics
confirmatory of our understanding of our data. Did R also give you the
correct degrees of freedom for error?

The t- and p-values provide tests of whether, for example the
gradient/slope is different from zero, and we can see that it is. In fact, we
might even report something like the following:

> Soil moisture had a positive effect on plant growth. For each unit increase in
> soil moisture, plant growth rate increased by 12.7 mm/week (slope = 12.7,
> $t = 12.5$, d.f. = 48, $p < 0.001$).

In case you were wondering, it is also be fine to report the *F*-value, degrees of freedom, and the *p*-value from **anova**(). They test exactly the same thing in this example. This isn't generally true of general linear models, though.

5.5.6 FROM STATS BACK TO FIGURE

The final step in our workflow involves translating the model we have fitted back onto our figure of the raw data. In this simple linear regression example, *ggplot2* comes to the rescue. For more complicated models, we need to do something different (see Chapters 6 and 7), but for now, let's introduce one more feature of *ggplot2*: the capacity to add regression lines (Figure 5.6):

```
ggplot(plant_gr, aes(x = soil.moisture.content,
        y = plant.growth.rate)) +
  geom_point() +
  geom_smooth(method = 'lm') +
  ylab("Plant Growth Rate (mm/week)") +
  theme_bw()
```

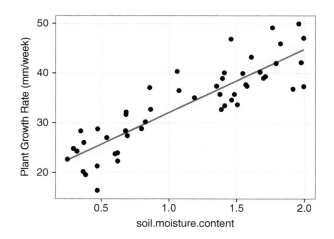

Figure 5.6 *ggplot2* provides the functionality to add the fitted values and standard error of the fit to a figure when it is a simple one-explanatory-variable model.

The extra, new feature we've introduced is the layer **geom_smooth**(method = "lm"), which is ***ggplot2***-speak for 'shove a linear-model fitted line, and the standard error of the fit, using flash transparent grey, onto my graph'. It is lovely and easy, isn't it?

Top cautionary tips. This is very handy. And the **geom_smooth**(method = "lm") tool is outstanding for this very simple example of a single explanatory variable. It is also outstanding for data exploration–i.e. *before* you make models, as it works brilliantly with **facet_wrap**(). But ... don't expect it to work correctly with more complicated models. We will, as promised, introduce the safe, robust method for adding model fits to your figure. Be patient ... we know you almost cannot wait.

OK. So, we've introduced, via a simple linear regression, some very classic tools for making inferences. We learned how to fit a linear model with **lm**(). We looked at four plots (produced by the **autoplot**() function) that tell us about how well model assumptions are met. We learned how to assess what **lm**() does using **anova**(), which doesn't fit an ANOVA, and **summary**(), which returns our estimates of the slope/gradient and intercept. Phew. And one more ***ggplot2*** trick. Nice.

Are you ready for one more example? Sure you are. Take a break if you need. We did ...

5.6 Analysis of variance: the one-way ANOVA

The final example in this chapter is a one-way ANOVA. The one-way ANOVA is as simple as the previous example, but we will see one change in the variables in the data frame: the explanatory variable is no longer continuous. It is a factor, or categorical variable.

You've experienced these types of variables already, using the compensation.csv data frame and, specifically, the Grazing variable, which had two levels, Grazed and Ungrazed. We continue that idea here, but with more levels. The dataset we are going to use centres on water fleas, also known more 'scientifically' as *Daphnia* spp., and their parasites, which have lots of cool names.

5.6.1 GETTING AND PLOTTING THE DATA

The question we are asking focuses on water flea growth rates and has two parts. First, we are asking generally whether parasites alter growth rates. Second, because it is a well-replicated and designed experiment, we can also ask whether each of three parasites reduces growth, compared with a control, no parasite treatment. We are going to use the same *Plot ->Model -> Check Assumptions -> Interpret -> Plot Again* workflow. Let's get started. Grab the `Daphniagrowth.csv` data from the same place as you get all the datasets, set up your script, make **dplyr**, **ggplot2**, and **ggfortify** available, clear the decks and check out the data... phew. This step is a lot easier now, isn't it? We are going to call the data *daphnia*:

```
glimpse(daphnia)
```

```
## Observations: 40
## Variables: 3
## $ parasite    (fctr) control, control, control, control...
## $ rep         (int) 1, 2, 3, 4, 5, 6, 7, 8, 9, 10, 1, 2...
## $ growth.rate (dbl) 1.0747092, 1.2659016, 1.3151563, 1....
```

We can see that the data frame has three variables, two of which we want to use for figure making, **growth.rate** and **parasite**. The other, **rep**, indicates the replication in each treatment level. Making a figure of these data needs a bit of thinking before jumping in. Recall Chapter 4, where we introduced the box-and-whisker plot as a quick and effective tool for viewing variation in a response variable as a function of a grouping, categorical variable? That's probably the start we want to make (Figure 5.7):

```
ggplot(daphnia, aes(x = parasite, y = growth.rate)) +
  geom_boxplot() +
  theme_bw()
```

This looks excellent, and it's just what we expected. Save for the fact that the parasite names are *huge* and mashed together like sardines. *Not. Good. Need. Quick. Fix.* Chapter 8 is going to provide a series of extra tools to manipulate the axis attributes. We are going to introduce here a more bizarre,

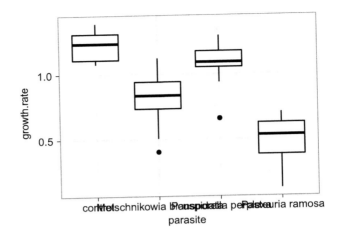

Figure 5.7   Daphnia's parasites alter daphnia growth rate.

and fun, trick. It's called coordinate flipping. Having made the graph in a way that works perfectly for **geom_boxplot()**, we just are going to switch the *x*- and *y*-axes using **coord_flip()**. This may seem novel to some of you, but it can be a very good way to visualize categorical data:

```
ggplot(daphnia, aes(x = parasite, y = growth.rate)) +
  geom_boxplot() +
  theme_bw() +
  coord_flip()
```

That's better. It's not perfect, but we can certainly read the names now AND we can start the process of second-guessing the statistics!

The first thing we can note is that there is substantial variation in the daphnia growth rates among the four treatments. Second, we can see that the control treatment produces the highest growth rate, about 1.2 mm/day. Third, we can see that *P. perplexa* is closest and perhaps lower, *M. bicuspidata* (try saying that first name!) is next lowest, and *P. ramosa* is definitely lower. Or we believe so. The point is, we can see that there is likely to be a parasite treatment effect overall (question 1), and an ordering in the growth rates, with parasites generally driving down the growth rate, and *P. ramosa* < *M. bicuspidata* < *P. perplexa* (question 2).

You can go even further if you wish, estimating the average growth rate for each treatment (looks like about 1.2 for the control treatment), and the difference of the parasite treatments from the control treatment (i.e. the effects of each parasite on growth rate). You could also figure out the degrees of freedom for the treatment and the degrees of freedom for error. It's very good practice to do this now, and to always do it, and then check R reports what you expect.

With that in mind, let's construct our linear model that does a one-way ANOVA. Don't freak out. The function that performs a one-way ANOVA is... lm().

5.6.2 CONSTRUCT THE ANOVA

This should be familiar. It has exactly the same structure as the linear regression model, except that 'we know' that the explanatory variable is a categorical variable:

```
model_grow <- lm(growth.rate ~ parasite, data = daphnia)
```

5.6.3 CHECK THE ASSUMPTIONS

That was, like, the shortest section ever. And here we are again, ready to evaluate the set of assumptions associated with linear models. They are the

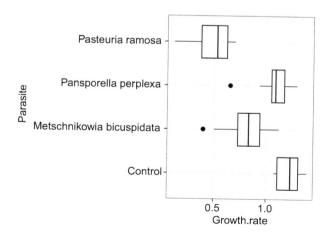

Figure 5.8 Perhaps a nicer graph of how daphnia's parasites alter daphnia growth rate.

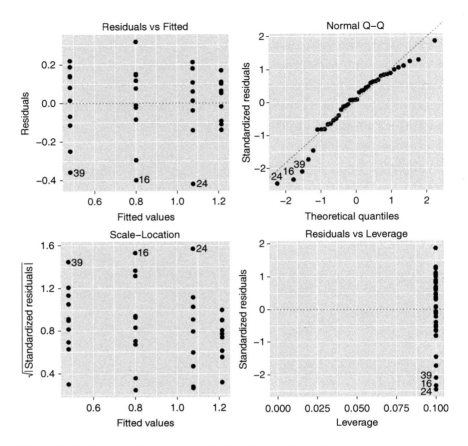

Figure 5.9 Nothing looks too bad in the diagnostic plots.

same for a regression and for a one-way ANOVA. We construct them and evaluate them in exactly the same way as above, by plotting four diagnostic graphs (Figure 5.9):

```
autoplot(model_grow, smooth.colour = NA)
```

If this didn't work, then you've not added **library**(ggfortify) to your script ...

Briefly, these figures suggest everything is probably fine. You may be uncomfortable with the Q–Q plot, upper right, evaluating the normality of the residuals. You can trust us that this pattern is within the expected bounds of variation found in samples of the normal distribution. If there is a 'real' departure from normality, it isn't too excessive, so our all-important p-values should be OK. Or, you could read more, simulate stuff, and prove to yourself that it's OK.

We are going to consider them fine, and move on to interpreting the output of the two tools for making inferences from linear models, **anova**() and **summary**(). Yes, we are going to apply the function **anova**() to a one-way ANOVA.

5.6.4 MAKING AN INFERENCE FROM A ONE-WAY ANOVA

As above, we'll start with the use of **anova**(). This function, applied to our model, will provide an answer to our first question posed above: is there an effect at all of our treatments?

```
anova(model_grow)

## Analysis of Variance Table
##
## Response: growth.rate
##
##            Df Sum Sq Mean Sq F value    Pr(>F)
## parasite    3 3.1379 1.04597  32.325 2.571e-10 ***
## Residuals  36 1.1649 0.03236
## ---
## Signif. codes:
## 0 '***' 0.001 '**' 0.01 '*' 0.05 '.' 0.1 ' ' 1
```

The output looks remarkably similar to that from a regression, and it should. They are both just linear models. We can see that there is indeed evidence that the parasite treatment, comprising four levels of manipulation, has produced an effect.

It is worth considering exactly what the null hypothesis is for a one-way ANOVA: that all of the groups come from populations with the same mean. The F-value quantifies the ratio of the between-group variance to the within-group variance. As the former is large relative to the latter, this produces a large F-value, and thus a small p-value that allows us to reject the null hypothesis that there are no differences.

Moving along to our second question, what are the effects? This is a question that can be answered in many ways. We'll show you one, and discuss a few others. The one we'll show you involves understanding how R presents coefficients of linear models with categorical explanatory variables.

5.6.5 TREATMENT CONTRASTS

Every statistical package has to decide how to produce the information in a summary-like table. There are many ways to do this, and these 'ways' are known as contrasts. R uses a presentation method known as 'treatment contrasts'. Several other statistical programs do as well, including Genstat and ASREML. SAS does not. Minitab does not. This means that if you compare the outputs of various statistical programs, they may well be different. None are wrong. Just different.

If you don't know what contrasts are or mean, they are a way of expressing coefficients taken from statistical models, and there are many types; see Crawley (2012) and Venables and Ripley (2003). Understanding what type of contrast is used by a statistical package is required for interpreting the summary table. This is also important when you make comparisons between the results from R and other packages. We suggest that those of you who are transitioning from other packages take a good look at ?contr.treatment, as this help file will explain R's contrasts and also reveal methods for making R produce output that matches other statistical packages.

Let's learn what R does, and reveal how a bit of luck and these treatment contrasts give us the answer to our second question. Lets begin by getting the summary table:

```
summary(model_grow)

##
## Call:
## lm(formula = growth.rate ~ parasite, data = daphnia)
##
## Residuals:
##      Min      1Q   Median      3Q      Max
```

```
## -0.41930 -0.09696  0.01408  0.12267  0.31790
##
## Coefficients:
##
##                                     Estimate Std. Error
## (Intercept)                          1.21391    0.05688
## parasiteMetschnikowia bicuspidata   -0.41275    0.08045
## parasitePansporella perplexa        -0.13755    0.08045
## parasitePasteuria ramosa            -0.73171    0.08045
##
##                                     t value Pr(>|t|)
## (Intercept)                          21.340  < 2e-16 ***
## parasiteMetschnikowia bicuspidata    -5.131 1.01e-05 ***
## parasitePansporella perplexa         -1.710   0.0959 .
## parasitePasteuria ramosa             -9.096 7.34e-11 ***
## ---
## Signif. codes:
## 0 '***' 0.001 '**' 0.01 '*' 0.05 '.' 0.1 ' ' 1
##
## Residual standard error: 0.1799 on 36 degrees of freedom
## Multiple R-squared:  0.7293, Adjusted R-squared:  0.7067
## F-statistic: 32.33 on 3 and 36 DF,  p-value: 2.571e-10
```

Let's start slowly. Notice that there are four rows in the 'table' of co-efficients (estimates), and the first row is labelled '(Intercept)'. If you look along that row, you may see a number you recognize: 1.2. Hmm. Interesting. Let's come back to that. Below that are the names of the three parasites. Hmm. Interesting. What's missing? The level of the treatment group called 'control'. But it's not missing. It is labelled '(Intercept)'.

Top Tip with ANOVA. The most important thing to get your head around with 'treatment contrasts' is the alphabet. R just loves the alphabet. In fact, it defaults to presenting things in alphabetical order. In this example, if we look at the alphabetical order of all of the treatment levels, we find that *control < M. bicuspidata < P. perplexa < P. ramosa*. This is the order presented in the table obtained from **summary()** and in the figure. Consistency is good.

So, in an ANOVA framework, we can assume that the word '(Intercept)' represents the first level of the alphabetically ordered treatment levels. We hope that makes sense. If it does, then you are ready for the fun bit.

Treatment contrasts report *differences* between the reference level (in this lucky case, the control) and the other levels. So, in the summary table, the numbers associated with each parasite are differences between growth

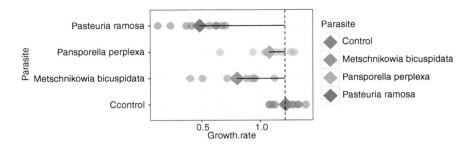

Figure 5.10   Raw growth rates, mean growth rates, and differences between control and parasite manipulations.

rates associated with that parasite and the control. This is why they are all negative. These negative distances, or contrasts, are the lengths of the black lines in Figure 5.10.

You can also get the means for each treatment level/group using *dplyr*, **group_by**(), and **summarise**(), and then calculate the contrasts yourself:

```
# get the mean growth rates
sumDat<-daphnia %>%
  group_by(parasite) %>%
    summarise(meanGR = mean(growth.rate))

sumDat

## Source: local data frame [4 x 2]
##
##                    parasite    meanGR
##                      (fctr)     (dbl)
## 1                   control 1.2139088
## 2 Metschnikowia bicuspidata 0.8011541
## 3       Pansporella perplexa 1.0763551
## 4           Pasteuria ramosa 0.4822030
```

Now you can calculate by hand how 1.21 (the control growth rate) plus the treatment contrast for a given parasite in the **summary**() gives you the mean growth rate for that parasite. For example, using the *P. ramosa* coefficient (see the summary table above), we take 1.21 – 0.73 = 0.48, which is the mean growth rate for *P. ramosa*. Remember the alphabetical rule, know your reference level, and then watch the numbers materialize.

Back to that summary table one last time. We said we were lucky because the control ended up as the reference group. This means the p-values associated with the contrasts are actually useful. They tell us whether the difference between the growth rate for parasite treatment and for the control is significant, i.e. they allow us to ask whether a particular parasite is reducing the daphnia growth rate. However, if you've ever heard of the 'multiple testing problem', you might be inclined to be careful with those p-values. Ask a knowledgeable friend for a little guidance if that means nothing to you.

Finally, you may want to produce a figure, something like what we've done above. Aside from the segments, this is rather straightforward, and you can probably figure it out. Note how we've specified two **geom_point()** layers? The first inherits the aesthetics from the daphnia data frame. The second, however, specifies the sumDat data frame we made above, using that as a source of data to plot the mean growth rates, with a big diamond (Figure 5.11).

In the next chapter we cover what to do if you're not so lucky with the word 'control' being the first in the alphabetical listing of the treatment levels. This can be a problem, since it will mean the coefficients reported in the summary table are not so useful, as they won't be differences of treatments from control. Spoiler, the solution involves the **relevel()** function.

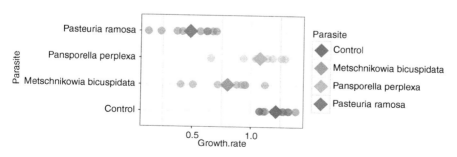

Figure 5.11 Raw growth rates and mean growth rates for each parasite manipulation.

5.7 Wrapping up

The take-home messages from this chapter are:

- Have a consistent workflow for analyses.
- Always make a graph that answers your question before touching any statistical tests.
- Interpret as much as you can, for example slopes, intercepts, contrasts, and degrees of freedom, from your graph *before doing any statistics*.
- Check if your data violate any assumptions, *before* interpreting your model results.
- Make a beautiful graph to communicate your results, and write a biologically focused (rather than statistically focused) sentence describing your results.
- R makes all this very easy.

Appendix Getting packages not on CRAN

There may be times when a package such as ***ggfortify*** may not be available on CRAN; perhaps it has not passed some test, or is having a bad day. For example, during the writing of this book, ***ggfortify*** was unavailable for a while. It had been removed from CRAN for some reason, and when we tried to install it, we got the warning shown in Figure 5.12, and the package didn't install. If this happens, you have a couple of options.

First, if the package was once on CRAN and was archived, an older version may still be available. You can probably find this by googling 'ggfortify cran' and following the links to a rather unfriendly-looking

```
> install.packages("ggfortify")
Warning in install.packages :
  package 'ggfortify' is not available (for R version 3.2.4)
```

Figure 5.12 A warning message that once appeared when we tried to install the ***ggfortify*** package.

web page that lists a few files, one of which was, at the time of writing, `gfortify_0.1.0.tar.gz`. Download this to your computer, making sure you know which folder it gets downloaded into. Then, in RStudio, click the *Install* button in *Packages* (as usual when you want to install a package). However, instead of using *Install from: Repository CRAN* choose *Install from: Package Archive File (.tgz, .tar.gz)*. Then you need to click *Browse* and find the file you just downloaded. Then click *Install*. Hopefully it will still work with the version of R you have installed.

The second option is perhaps rather simpler. Many packages are now developed in an environment/web service called GitHub, and R has an interface for accessing packages in developmental stages directly from GitHub. This is true for **ggfortify**. Here is how you can access the package that way:

```
install.packages("devtools")
library(devtools)
install_github("sinhrks/ggfortify")
```

First you get and load a new library, **devtools**. Then you install the package you want from a website called `github`. (`sinhrks` is the GitHub username of the package maintainer and co-author, Masaaki Horikoshi.) This solution will probably give you a development version of the package, rather than the released one. This probably won't matter; you won't notice any difference.

6

Advancing Your Statistics in R

6.1 Getting started with more advanced statistics

Chapter 5 was pretty intense. We introduced statistics and interpretation of R output for four different examples: a χ^2 contingency table analysis, a *t*-test, a simple linear regression, and a one-way ANOVA. In this chapter we extend the complexity of our models to introduce analyses with more than one explanatory variable. In fact, we focus on models with two explanatory variables: the two-way ANOVA and the ANCOVA. Our emphasis, as usual, remains on how to accomplish these in R, rather than on the details of the statistical methods. As we said in Chapter 5, if you're not familiar with the statistical foundations of these tools, you should take the time to become so. Our goal is to introduce to you the formulation and interpretation of such models in R. And nothing changes about our workflow of *Plot -> Model -> Check Assumptions -> Interpret -> Plot Again*. Except that the benefits of **dplyr** and **ggplot2** may become more apparent.

6.2 The two-way ANOVA

Chapter 5 finished with the one-way ANOVA, and the major feature of this model was that the explanatory variable was categorical. Recall that

Getting Started with R Second Edition. Andrew Beckerman, Dylan Childs, & Owen Petchey:
Oxford University Press (2017). © Andrew Beckerman, Dylan Childs, & Owen Petchey.
DOI 10.1093/oso/9780198787839.001.0001

the explanatory variable was **parasite** and it had four levels. The two-way ANOVA is a logical extension of this, and involves experiments or structured data collection where two explanatory variables are involved. As with the one-way ANOVA, the explanatory variables are, in a two-way ANOVA, both categorical.

The two-way ANOVA thus corresponds to data that have structure in two dimensions. The response variable may vary with both variables. In fact, the most exciting and motivating reason for designing experiments and data collection associated with a two-way ANOVA analysis is that the way in which the response variable varies with one variable may depend on the other variable. This is a statistical interaction. The idea of an interaction is often central to hypotheses involving more than one explanatory variable. The word 'depends' is a great one to remember as we move on in the book.

6.2.1 THE COW (MOOOOO!) GROWTH DATA

Let's get some example data and work through its structure to articulate the nature of a hypothesis associated with a two-way ANOVA. We are going to work with growth data for cows. The cows were fed one of three diets: barley, oats, and wheat (do you notice the alphabetical order?). The diets were also enhanced with one of four supplements (control and some funny names again). The important thing is that the data are associated with a fully factorial experimental design, where each combination of diet and supplement was replicated three times. This means that there are three diets × four supplements = 12 treatment combinations, each with three cows in them (mooooo! 36 times).

Now, you can (and should) work out the degrees of freedom for the error, each 'main effect', and 'the interaction'. If you have no idea what all the key phrases in that sentence mean, now is a very good time to do a little more reading. Let's see what we can also glean from a picture of the data.

The name of the data file is growth.csv and it is, like the others, available from http://www.r4all.org/the-book/datasets/.

Set up a script, and, as usual, *Annotate, clear the decks, make libraries avail-able, and import the data, followed by, for example, glimpse().* We'll call the data growth.moo.

Check the data was imported correctly, and check their structure:

```
glimpse(growth.moo)

## Observations: 48
## Variables: 3
## $ supplement (fctr) supergain, supergain, supergain, su...
## $ diet       (fctr) wheat, wheat, wheat, wheat, wheat, ...
## $ gain       (dbl) 17.37125, 16.81489, 18.08184, 15.781...
```

As we expect for a two-way ANOVA, two of the variables are factors and one is labelled 'dbl', indicating numeric. This is to plan. We'll introduce a cool function from *base* R here, called **levels()**. This function allows us to see the levels associated with a factor. It's very handy. We use **levels()** with '$', grabbing the column from which we want the information:

```
levels(growth.moo$diet)

## [1] "barley" "oats"    "wheat"
levels(growth.moo$supplement)

## [1] "agrimore" "control"   "supergain" "supersupp"
```

Pay attention to the order of things. One of the things we can immediately see is that the supplements include a control level, but the agrimore level is in front of it alphabetically. Remember the reference level? We probably want the control to be the reference level. Here we employ that nifty function **relevel()**, which, for just this purpose, creates a new reference level by re-ordering the factor. We use it with the **mutate()** function from *dplyr* as follows:

```
# relevel the supplement column
growth.moo <-
  mutate(growth.moo,
         supplement = relevel(supplement, ref="control"))

# check it worked
levels(growth.moo$supplement)

## [1] "control"   "agrimore"   "supergain" "supersupp"
```

How cool! We've overwritten the supplement column in the imported data, forcing the control to be the reference. This is good. Can you guess what level of diet is the reference? That's right... Oats.

Now we are ready to make a fancy figure of the data. This is quite fun. We'll use **dplyr** to calculate the mean of the growth rates for each of the 12 combinations, and then use **ggplot()** to plot them. What we are producing is an elegant interaction plot. So, first, use **dplyr** to get means. Second, get these summary statistics onto an informative plot. Let's go.

6.2.2 STEP 1: *dplyr* SUMMARY DATA

Let's stick with the piping syntax and recall that our data frame is called growth.moo. Let's also note that we have two grouping variables, both of which we need to give to the **group_by()** function:

```
# calculate mean and sd of gain for all 12 combinations
sumMoo <- growth.moo %>%
  group_by(diet, supplement) %>%
    summarise(meanGrow = mean(gain))

# make sure it worked
sumMoo

## Source: local data frame [12 x 3]
## Groups: diet [?]
##
##       diet supplement meanGrow
##      (fctr)     (fctr)    (dbl)
## 1  barley     control 23.29665
## 2  barley    agrimore 26.34848
## 3  barley   supergain 22.46612
## 4  barley   supersupp 25.57530
## 5    oats     control 20.49366
## 6    oats    agrimore 23.29838
## 7    oats   supergain 19.66300
## 8    oats   supersupp 21.86023
## 9   wheat     control 17.40552
## 10  wheat    agrimore 19.63907
## 11  wheat   supergain 17.01243
## 12  wheat   supersupp 19.66834
```

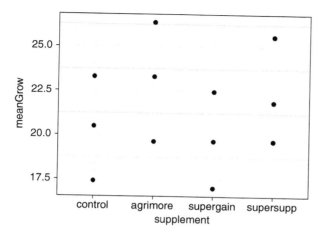

Figure 6.1 Cow weight gain, unadorned.

6.2.3 STEP 2: **ggplot()** INTERACTION PLOT

OK. Now for the fun bit. We are going to first add the points from sumMoo to a figure (Figure 6.1), and specify that the *x*-axis is supplement. The logic of this comes from thinking about this statement, to which we will return:

> The effect of supplement type on cow weight gain depends on the diet.

```
ggplot(sumMoo, aes(x = supplement, y = meanGrow)) +
   geom_point() +
   theme_bw()
```

OK. The next step in this figure-making process involves connecting the dots (Figure 6.2). Our objective is to connect them according to diet, and perhaps to ask for colours to correspond to diet. We do this in the original aesthetics, specifying both colour = diet *and* group = diet, and by asking for another layer, of lines:

```
ggplot(sumMoo, aes(x = supplement, y = meanGrow,
                   colour = diet, group = diet)) +
   geom_point() +
   geom_line() +
   theme_bw()
```

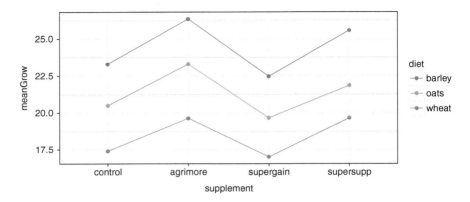

Figure 6.2   Does the effect of supplement type on cow weight gain depend on the diet?

6.2.4 INTERPRETING THE FIGURE: BIOLOGICAL INSIGHT

Sweet mother of cows, milk! What might we conclude from Figure 6.2? Do you think that the effect of **supplement** type on cow weight **gain** depends on the **diet**? One way to think about this is to first look at the pattern that the **supplements** generate for the oat **diet**. Then, ask yourself, does this pattern differ for the other **diets**, or is it the same? Geometrically, are the effects of **supplement** generating parallel patterns among the **diets**, or not? If your answer is that the patterns are the same and the lines are roughly parallel, you have probably guessed that there is *no interaction.* The effect of **supplement** type on cow weight **gain** *does not* depend on the **diet**.

This does not mean that there is no effect of **diet**. In fact, we can see that, on average, if you are a cow, eating barley rocks the world. This also does not mean that there is no effect of **supplement**; two of them seem to improve growth no matter what the diet!

Together, the similarity of all the patterns, and the evidence that both variables seem to have an effect, suggests an additive rather than interactive set of effects. Grand. Now we are ready to build the model.

6.2.5 CONSTRUCTING THE TWO-WAY ANOVA

At this point, let us formally specify the null hypothesis we are testing. Wait! We have. But here we go again. The effect of supplement type on cow weight gain *does not* depend on the diet. We wish to emphasize that if one designs an experiment with two factors, often one *is* interested in the interaction—whether the effect of one variable on the response variable depends on a third. We note too that the null hypothesis is thus the 'additive' model. The alternative is that there is an interaction. All of this means that we want to fit a model *with* the interaction. As we'll see, this allows us to test formally between the additive and interaction (non-additive) alternatives.

To build the model, we use a trick common to most statistical programming languages. We use a * symbol between the two explanatory variables. Specifically, if we write diet * supplement, it will expand the combination of variables to read diet + supplement + diet:supplement, such that we have an a (main) effect of diet, a (main) effect of supplement, and the interaction (:) between diet and supplement.

As with all linear models, we use the **lm**() function, specifying this trick formula and the data frame:

```
model_cow <- lm(gain ~ diet * supplement, data = growth.moo)
```

We will come back later to showing you that the expansion of * worked. But first, we must check the assumptions! Which are the same as they were for the regression and the one-way ANOVA. So lets build those four plots:

```
autoplot(model_cow, smooth.colour = NA)
```

6.2.6 EXAMINE THE ASSUMPTIONS

These plots look OK (Figure 6.3), though they aren't perfect. The residuals vs fitted values (top left) do not show any kind of pattern to suggest our model is inappropriate. In fact, since we built a full model that includes the interaction (a very flexible model), this plot really should be OK. The

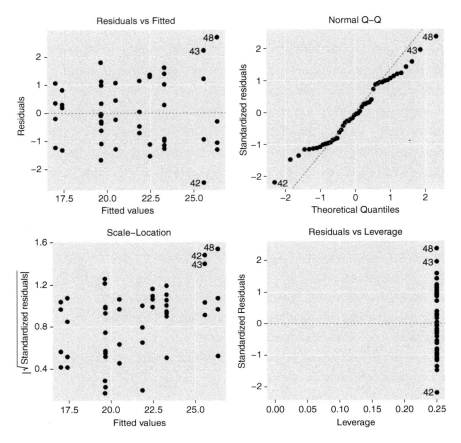

Figure 6.3 The diagnostic plots for the cow weight gain two-way ANOVA.

scale–location plot (bottom left) has almost no pattern, and as you'll see later in Chapter 7, what we see here is nothing to worry about. There also don't seem to be any serious outliers either (bottom right).

The normal Q–Q plot (top right) is not super-great. It isn't disastrous either. The positive and negative residuals tend to be smaller in magnitude than they should be. If you are curious, this arises because the tails (the 'ends') of the residuals' distribution are 'squashed' towards its middle, relative to what we expect. This is a rather unusual pattern. It is almost as though someone made up these data, and chose to do this in a strange way. . . The good news is that a departure from normality like this shouldn't mess up our stats *too* much. The general linear model is actually quite robust to moderate deviations from normality, so let's keep going. This is a book about using R, after all.

6.2.7 MODEL OUTPUT AND MORE BIOLOGY

Now we're ready for the fun and games. Remember, there are two functions to aid in interpretation of a general linear model: **anova()** and **summary()**. **anova()** produces an ANOVA table listing sums of squares, mean squares, *F*-values, and *p*-values. It *does not* perform an ANOVA. **summary()** does many things, as you might recall. When provided with a model object from **lm()** as its argument, **summary()** returns a table of coefficients (slopes and intercepts), standard errors, and *t*-values.

Let's start with **anova()**:

```
anova(model_cow)

## Analysis of Variance Table
##
## Response: gain
##
##                  Df  Sum Sq Mean Sq F value    Pr(>F)
## diet              2 287.171 143.586 83.5201 2.999e-14 ***
## supplement        3  91.881  30.627 17.8150 2.952e-07 ***
## diet:supplement   6   3.406   0.568  0.3302    0.9166
## Residuals        36  61.890   1.719
## ---
## Signif. codes:
## 0 '***' 0.001 '**' 0.01 '*' 0.05 '.' 0.1 ' ' 1
```

As we promised, using diet*supplement has indeed expanded to provide three rows, corresponding to the variation explained by diet, the supplement, and the interaction. Let's walk through the logic, biological and statistical, of this table.

The ANOVA table

The ANOVA table presents a sums-of-squares analysis-of-variance table. This is interpreted as follows. The first line shows the variation explained by diet, captured in the Mean Sq value of 143.586. Having captured this variation, we then see an estimate of the variation explained by the supplement, captured in the Mean Sq value of 30.627. Finally, having captured both of these estimates of variation, we finally ask whether any additional amount of variation is explained by allowing the effect of supplement to vary by diet. The answer is captured in a tiny (0.568) Mean Sq. The F-statistics for each of these parts of our explanation are calculated by dividing each Mean Sq by the residual Mean Sq (last line), generating two significant p-values for the main effects and a non-significant value for the interaction.

Because we conducted, and are analysing, an experiment, we had a specific hypothesis in mind—that the effect of **supplement** on weight **gain** depended on **diet**. Testing this hypothesis is embodied in the diet:supplement row. This table reveals that there is no additional, significant variation in weight **gain** explained by allowing the effect of **supplement** to vary by **diet** ($F = 0.33$; $df = 6, 36$; $p = 0.92$). Maybe this was obvious from the figure we made, but now we have the statistical evidence we need to support this insight. Here is how you might report this result:

> We tested the hypothesis that the effect of diet supplement on bovine weight gain depended on cereal diet. We found no evidence to support the presence of an interaction between diet and supplement ($F = 0.33$; $df = 6, 36$; $p = 0.92$).

Hopefully, the explanations along the way have been helpful. The ANOVA table is all about explaining variation associated with main effects and interactions.

We note that the order of the variables in the ANOVA table in this example does not matter, because the data are from a designed experiment where each of the explanatory variables are independent by design. The data are from a fully factorial, balanced experiment. If a dataset does not posess these properties (e.g. in most observational studies), then the order of testing in the ANOVA table *does* matter. It matters because the mean squares, F-values, and ultimately the p-values will depend on the order of testing. This is not the place to discuss this idea in detail, but we will repeat this important warning (it would be irresponsible not to): unless your data are balanced and orthogonal, the order of testing in an ANOVA table matters. This means you should not blindly rely on the output of **anova()** in these situations. There is a lot of good reading to do on this and we encourage you to seek some of it out if you're not sure what we're talking about (see Appendix 2 and, in particular, the treatment in *An R Companion to Applied Regression* by J. Fox).

We could actually stop the analysis here. We have analysed data from an experiment designed to test formally a hypothesis about whether the effect of supplements on cow weight gain depends on the diet. R has served us well, and provided an answer.

The summary table

However, we may want more. Let's not forget that there is a summary table. This table is quite large. Let's look at it and then make some recommendations about how to proceed, if one feels compelled:

```
summary(model_cow)

## 
## Call:
## lm(formula = gain ~ diet * supplement, data = growth.moo)
## 
## Residuals:
##      Min      1Q   Median      3Q     Max
## -2.48756 -1.00368 -0.07452  1.03496  2.68069
## 
## Coefficients:
##                                  Estimate Std. Error t value
## (Intercept)                    23.2966499  0.6555863  35.536
## dietoats                       -2.8029851  0.9271390  -3.023
## dietwheat                      -5.8911317  0.9271390  -6.354
## supplementagrimore              3.0518277  0.9271390   3.292
## supplementsupergain            -0.8305263  0.9271390  -0.896
## supplementsupersupp             2.2786527  0.9271390   2.458
## dietoats:supplementagrimore    -0.2471088  1.3111726  -0.188
## dietwheat:supplementagrimore   -0.8182729  1.3111726  -0.624
## dietoats:supplementsupergain   -0.0001351  1.3111726   0.000
## dietwheat:supplementsupergain   0.4374395  1.3111726   0.334
## dietoats:supplementsupersupp   -0.9120830  1.3111726  -0.696
## dietwheat:supplementsupersupp  -0.0158299  1.3111726  -0.012
##                                Pr(>|t|)
## (Intercept)                     < 2e-16 ***
## dietoats                        0.00459 **
```

```
## dietwheat                        2.34e-07 ***
## supplementagrimore               0.00224 **
## supplementsupergain              0.37631
## supplementsupersupp              0.01893 *
## dietoats:supplementagrimore      0.85157
## dietwheat:supplementagrimore     0.53651
## dietoats:supplementsupergain     0.99992
## dietwheat:supplementsupergain    0.74060
## dietoats:supplementsupersupp     0.49113
## dietwheat:supplementsupersupp    0.99043
## ---
## Signif. codes:
## 0 '***' 0.001 '**' 0.01 '*' 0.05 '.' 0.1 ' ' 1
##
## Residual standard error: 1.311 on 36 degrees of freedom
## Multiple R-squared:  0.8607, Adjusted R-squared:  0.8182
## F-statistic: 20.22 on 11 and 36 DF,  p-value: 3.295e-12
```

The table looks rather scary. We argue that in fact this table, as is, is probably not what you want. It is in fact interpretable. We've set you up to understand at least parts of it. Recall our discussion of treatment contrasts in Chapter 5. And recall above our emphasis on alphabetical ordering of both explanatory variables. Together, these two points should help you to interpret at least the first three lines.

The '(Intercept)' is a reference point for the table. And it is a combination of levels, one for each variable. Can you figure it out? It has to be barley–control because barley is the first **diet** in the alphabetized list, and we used **relevel**() to force Control to be the reference level. You can find the value 23.29 on the graph, in fact in the top left.

Great. Now, recall that all the other estimates are differences between this reference level and whatever is labelled in the row. So, rows two and three specify the oats and wheat, indicating that these values are the differences between barley–control and oat–control, and between barley–control and wheat–control, respectively. Again, you can prove this to yourself with the numbers and the graph.

But, is this what you really want to do? We think not. In fact, we think that if you had designed such a large, multi-factorial experiment, you probably had some *specific, a priori* hypotheses in mind (we really hope you did). In stats speak, a priori hypotheses are typically described as *contrasts*. A contrast is just a difference between the means of two levels, or some combinations of levels. You know about treatment contrasts, but you can use R to examine the statistical significance of any contrast you like (even nonsensical ones).

You can, in fact, specify contrasts before you fit the model, and just estimate the subset for which you designed the experiment. This. Is. Very. Good. Practice. We can recommend several packages to help you work with contrasts. The **contrast**, **rms**, and **multcomp** packages all contain excellent facilities for specifying custom contrasts. Furthermore, the **multcomp** package has numerous built-in post-hoc (e.g. Tukey) tests that you may be interested in learning how to use. We think a priori contrasts are much better. This stuff is a bit tricky, though, so it is probably another find-an-experienced-friend scenario.

6.2.8 STATS BACK TO GRAPHICS

One might be tempted to conclude that the figure of the mean weight gain taht we've already made is sufficient to understand the cow diet data. However, we really need to understand the variation around each mean value. There are several ways to do this, but we'll take this opportunity to show you a bit more **dplyr** and **ggplot** loveliness.

First, we return to the use of **dplyr**, where we calculated meanGain. We want to also include an estimate of the standard error. We do this via the following code:

```
# calculate mean and sd of gain for all 12 combinations
sumMoo <- growth.moo %>%
  group_by(diet, supplement) %>%
    summarise(
      meanGrow = mean(gain),
      seGrow = sd(gain)/sqrt(n())
    )
```

Note that the standard error of the mean is given by the standard deviation divided by the square root of the sample size, and **dplyr** provides a function n() that counts the rows in each group. Be careful with this, however. If you have missing values, n() will still count these rows.[1]

---

[1] Here is an advanced tool to alleviate this problem: **sd**(gain)/**sqrt**(**sum**(!**is.na**(gain)))... check the help files and interweb!

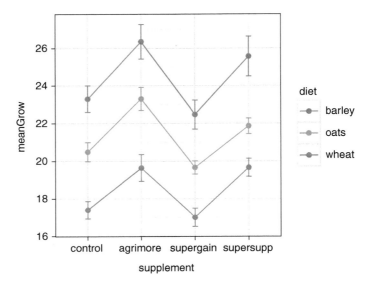

Figure 6.4   A quite nice graph of the cow growth data, finishing off our two-way ANOVA example.

With this summary table, we can now add three layers to the fig-ure: points, lines, and error bars (Figure 6.4). To do this, we introduce **geom_errorbar()**, which has its own aesthetics: the lower and upper limits of the vertical lines that we call error bars. These limits are called ymin and ymax. Each vertical line is made by placing a line between these limits, passing through meanGrow. The magic here is that we've got 12 estimates of mean weight gain and 12 estimates of the standard error. ***ggplot2*** is go-ing to add them ALL at once, correctly on each point. Note we also ask for the hat on the error bar to be small, using width. Did you know that error bars wear hats? Superb:

```
ggplot(sumMoo, aes(x = supplement, y = meanGrow,
                   colour = diet, group = diet)) +
  geom_point() +
  geom_line() +
  geom_errorbar(aes(ymin = meanGrow - seGrow,
                    ymax = meanGrow + seGrow), width = 0.1) +
  theme_bw()
```

6.3 Analysis of covariance (ANCOVA)

We hope you've taken a deep breath. And/or had a few more bis-cuits/cookies. Here we go for our final linear-model example. It is unique in that it combines a categorical explanatory variable with a continuous ex-planatory variable. What are we up to? We are combining regression and one-way ANOVA! Yes we are.

The dataset we will use is `limpet.csv` and is originally from Quinn and Keough's 2002 book *Experimental Design and Data Analysis for Biologists*. As usual, the data file is available at `http://www.r4all.org/the-book/datasets/`. The data relate egg production by limpets to four density conditions in two seasons. The response variable (*y*) is egg production (EGGS) and the independent variables (*x*'s) are DENSITY (continuous) and SEASON (categorical). Because we are examining egg production along a continuous density gradient, this is essentially a study of density-dependent reproduction.[2] The experimental manipulation of density was implemented in spring and in summer. Thus, a motivation for collecting these data could be 'does the density dependence of egg production differ between spring and summer'?

6.3.1 THE LIMPET REPRODUCTION DATA

As with all of our examples, set up a new script. We'll refrain from repeat-ing oursleves about what you should do... you know it already! We'll call the data frame 'limp':

```
glimpse(limp)

## Observations: 24
## Variables: 3
## $ DENSITY (int) 8, 8, 8, 8, 8, 8, 15, 15, 15, 15, 15, 1...
## $ SEASON  (fctr) spring, spring, spring, summer, summer...
## $ EGGS    (dbl) 2.875, 2.625, 1.750, 2.125, 1.500, 1.87...
```

---

[2] If you don't come from an ecology background, density dependence is the idea that as the number of mums sharing a food resource increases (e.g. their density goes up), they get a smaller and smaller portion, leading to reduced baby production.

6.3.2 ALWAYS START WITH A PICTURE

Let's begin the analysis as we normally do—with a fabulous picture in R.

You should notice that the dataset contains three columns, two that are numeric (EGGS and DENSITY) and a third that is categorical (SEASON). The response (*y*) variable is egg production, DENSITY is the continuous independent variable (*x*), and the values are also classified by the category SEASON. This suggests that a plot of egg production against density, distinguishing the two seasonal categories, is what we need. This is readily achieved in **ggplot**(), isn't it! The code has the same structure as what we introduced in Chapter 4. Make sure you understand how we've used **scale_colour_manual**() and how we've made sure that, oh, we can't resist, spring comes before summer! Ha!

```
# plot window
ggplot(limp, aes(x = DENSITY, y = EGGS, colour = SEASON)) +
  geom_point() +
  scale_color_manual(values = c(spring="green", summer="red")) +
  theme_bw()
```

Let's consider the pattern that we see in this graph (Figure 6.5). There is clearly a decline in egg production with increasing limpet density, and

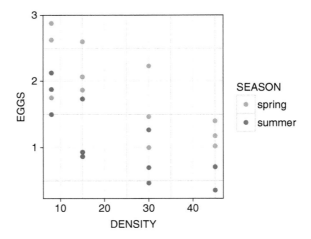

Figure 6.5 Always begin analyses with a picture. Our analysis of egg production by limpets in two seasons and at four limpet densities.

it looks as though, for any particular density, there is a tendency for the production to be higher in spring than in summer. So, just from the picture, we have extracted some pretty useful information. Now the challenge is how to test this more formally; in other words, how to make a model for the data.

To think about how to do this, we can start by recalling the equation for a straight line: $y = b + m \times x$. On our graph, if we were to describe the relationship between egg production and density by a straight line (ignore the season for the moment), then y is egg production, x is density, b is where the line crosses the y-axis (i.e. the egg production at 0 density—admittedly an odd concept!), and m is the slope of the egg production density relationship. This slope represents the change in egg production per unit change in density—that is, the strength of density dependence. m and b are the parameters, or coefficients, of the line.

To work out the degrees of freedom for error, you need to observe that there were three observations at each of the four densities in each of the two seasons (i.e. Figure 6.5 has 24 data points). We have data for two different seasons, so don't forget we'll work with a model that is estimating coefficients for two lines. You should also work out guesses of the values of the two intercepts and two slopes (i.e. the four coefficients that are estimated when we make the model).

6.3.3 INTERPRETING THE FIGURE—IT'S ALL ABOUT LINES

With the understanding that all straight lines are composed of an intercept and a slope, we can reinterpret our figure, assigning biological meaning to intercepts and slopes. First, we can note that, overall, egg numbers decline with increasing density—that is, there is a negative slope. Second, we can begin to speculate that there is a seasonal difference—that is, that the intercept, the value of egg production at zero density, is different in each season.

Once we start to look at the data plotted on the figure this way, we immediately start to identify the patterns that stand out. But in order to think about testing these, and interpreting what they mean, it is important to step back and consider all the possible patterns that we might have found, and the different hypotheses, or interpretations, they represent. We can do this using the two concepts of slope and intercept of the lines describing thse data. Figure 6.6 shows the range of possibilities, and describes each in verbal terms, graphically, and as an equivalent model in R.

Looking at our data above (Figure 6.5), we can walk through the different hypotheses in Figure 6.6. We might ignore A and B—it seems clear from the data that there is at least a decline, on average, in egg production with density. However, the higher the variation in egg production, the less likely we are to detect such a trend as significant. C is clearly possible; again, more so if the variation in egg production in each season is high. D and E are compelling. In our data, if the slopes are indeed equivalent, scenario D would be the best explanation. However, if the slopes are different and the lines for each season cross (even if they are both negative), then scenario E is a compelling, competing hypothesis. Because we can't actually see whether there are different slopes and intercepts, we can use statistics.

Let's go a bit deeper

For a number of reasons, we'd like you to understand that the difference in models D and E is the presence (E) or absence (D) of one particular term: an interaction term that specifies that the slopes are different. This interaction is embodied in the following sentence: 'The effect of density on egg production depends on the season.'

Try looking at Figure 6.5 and saying this out loud: 'The effect of density on egg production depends on the season.' If we deconstruct this sentence, we can reveal the features of a line. 'The effect of density on egg production' means the value of the slope(s); 'depends on the season' indicates that these values may be different depending on the season. Of course, if the slope depends on the season, we must also be estimating the effect of the season, in other words estimating intercepts as well as slopes. In fact, what statistical modelling does is to ask whether our data justify specifying more than one intercept and more than one slope. Put another way, we are

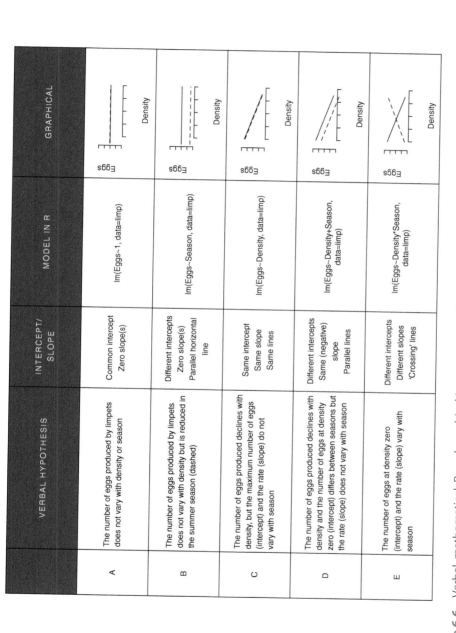

	VERBAL HYPOTHESIS	INTERCEPT/ SLOPE	MODEL IN R	GRAPHICAL
A	The number of eggs produced by limpets does not vary with density or season	Common intercept Zero slope(s)	lm(Eggs~1, data=limp)	
B	The number of eggs produced by limpets does not vary with density but is reduced in the summer season (dashed)	Different intercepts Zero slope(s) Parallel horizontal line	lm(Eggs~Season, data=limp)	
C	The number of eggs produced declines with density, but the maximum number of eggs (intercept) and the rate (slope) do not vary with season	Same intercept Same slope Same lines	lm(Eggs~Density, data=limp)	
D	The number of eggs produced declines with density and the number of eggs at density zero (intercept) differs between seasons but the rate (slope) does not vary with season	Different intercepts Same (negative) slope Parallel lines	lm(Eggs~Density+Season, data=limp)	
E	The number of eggs at density zero (intercept) and the rate (slope) vary with season	Different intercepts Different slopes 'Crossing' lines	lm(Eggs~Density*Season, data=limp)	

Figure 6.6 Verbal, mathematical, R, and graphical interpretations of various hypotheses related to the ANCOVA that translate into specific linear models.

asking: do we get a better description of the data by using more than one intercept and more than one slope?

At this stage, we have a good figure (Figure 6.5) that reveals a pattern and allows us to speculate about a result—some of you might think that there is a common rate of density dependence (slope), while others might see an interaction—there is the distinct possibility that the effect of density on egg production depends on the season.

Before we explore these alternatives, consider one more feature of these data: they were collected as part of a manipulative experiment. Philosophically, we may want to limit the number of competing hypotheses we consider (i.e. A–E) because when one designs an experiment, one usually has a particular hypothesis in mind; we have an a priori hypothesis. In the case of these data, if we assume that the researchers were testing whether 'The effect of density on egg production depends on the season', then in fact we are not equally interested in all the hypotheses: we are starting with the belief that density does have an effect on egg production, and we are focusing our investigation on whether this effect differs between seasons; that is, we are really interested in whether the data are better described by model E than by model D.

6.3.4 CONSTRUCTING THE ANCOVA

 To specify a general linear model (regression, ANOVA, ANCOVA), and one that would capture any of the above statistical hypotheses, we continue to rely on the function **lm**(), the workhorse of linear models in R:

```
limp.mod <- lm(EGGS ~ DENSITY * SEASON, data = limp)
```

We have assigned the model returned by **lm**() to the object `limp.mod`. And we have used a formula to specify the relationship between EGGS, DENSITY, and SEASON. As we discussed above, the formula's right-hand side expression is a sort of shorthand: it specifies, all at once, that we want to include an effect of DENSITY (main effect), an effect of SEASON (main effect), and the potential for the effect of DENSITY to depend on SEASON (interaction). The specification expands to the full model of DENSITY + SEASON + DENSITY : SEASON.

We could specify a model that doesn't include all these things—the other examples in Figure 6.6 illustrate what these would look like. But since we are interested in whether the effect of density depends on season, we need a model where there is, in addition to the effects of density and season on their own, a specific term allowing for their interaction, that is, in which the effect of density on egg production *could* depend on season.

Why do we say 'could'? We are, fundamentally, testing the null hypothesis (Figure 6.6, row E) that this interaction term is not significant: there are no differences in the slopes for each season; there is no extra variation explained by fitting different slopes. The alternative is that, by allowing separate slopes to be fitted, we explain more variation in the data. That's it. Said another way, we are asking if, having explained a certain amount of variation with different intercepts and a common slope (i.e. hypothesis D), do we explain a significant amount of additional variation by allowing separate slopes?

If you run the code above, all the good stuff has been done, and has been quietly collected and organized for you in the limp.mod object you specified.

As with all of the amazing R functions you have experienced thus far, by assigning the values returned by **lm()** to an object (limp.mod), we can see what has been collected for us; we've not done this for any of the previous examples, so let's have a look:

```
names(limp.mod)
```

```
##  [1] "coefficients"  "residuals"       "effects"
##  [4] "rank"          "fitted.values"  "assign"
##  [7] "qr"            "df.residual"    "contrasts"
## [10] "xlevels"       "call"            "terms"
## [13] "model"
```

R has stored for us the coefficients (intercept and slope estimates), the residuals, the fitted values, and a number of other values. Peruse the help file for **lm()**, specifically the Value section, to find out what **lm()** can return to you.

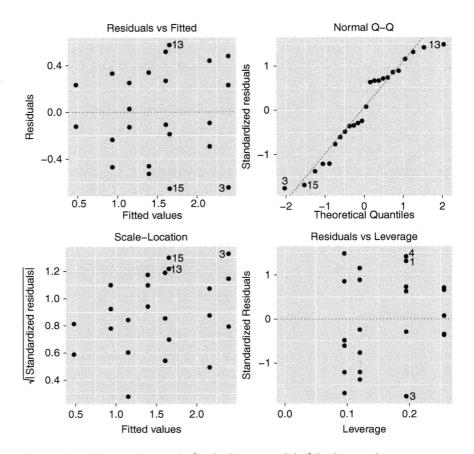

Figure 6.7   Diagnostic graphs for the linear model of the limpet data.

6.3.5 ASSUMPTIONS FIRST, SO SAY THE LIMPETS

By now you know what to do, right? Right.

```
autoplot(limp.mod, smooth.colour = NA)
```

The diagnostics are very nice (Figure 6.7). We saw some that weren't so great in the last example, and we'll see some more that are even worse in the next chapter.

We will assume that you, too, find the diagnostics satisfying, illuminating, and downright good. Before we head to our interpretation phase, let's review. We've made an insightful and useful figure representing the data, which helped us to formulate our ideas about how density and season might affect the egg production of limpets. We've made a model to capture our ideas, and, more importantly, an experimental design. And we've evaluated the core assumptions associated with a general linear model. Now you're ready for interpretation.

6.3.6 INTERPRETATION: THE anova() TABLE

As discussed throughout so far, we use two functions to aid in interpretation of a general linear model: **anova()** and **summary()**. Together, the resulting two tables help us interpret the relationships between egg production and the combined or independent impact of density and season. Sounds easy? If you know how to interpret the output of other packages, and understand contrasts and statistics deeply, it is. If you've never delved deeply into what a statistical package returns to you, hold on tight! In this example, the outputs both **anova()** and **summary()** are totally and immediately relevant.

Let's start with **anova()**:

```
anova(limp.mod)

## Analysis of Variance Table
##
## Response: EGGS
##
##                 Df Sum Sq Mean Sq F value    Pr(>F)
## DENSITY          1 5.0241  5.0241 30.1971 2.226e-05 ***
## SEASON           1 3.2502  3.2502 19.5350 0.0002637 ***
## DENSITY:SEASON   1 0.0118  0.0118  0.0711 0.7925333
## Residuals       20 3.3275  0.1664
## ---
## Signif. codes:
## 0 '***' 0.001 '**' 0.01 '*' 0.05 '.' 0.1 ' ' 1
```

Here we see again a sequential sums-of-squares analysis-of-variance table. This is interpreted as follows. First, we estimate a common slope (DENSITY), using 1 df, and this explains a certain amount of variation,

captured in the Mean Sq value of 5.0241. Having done this, we then estimate different intercepts for the model (SEASON) and, having done so,
explain an additional amount of variation (3.2502). Finally, we allow the
slopes to vary (DENSITY:SEASON), and this explains an additional 0.0118
units of variation. This is the sequence of explanation.

Because we conducted, and are analysing, an experiment, we had a
specific hypothesis in mind—that the effect of density on egg production depends on the season. Testing this hypothesis is embodied in the
DENSITY:SEASON row.

This table reveals that there is no additional, significant variation in egg
number explained by allowing different slopes for each season (i.e. the
p-value for the DENSITY:SEASON row is 0.79). We fail to reject the null
hypothesis that the slopes are the same, i.e. we have failed to find evidence
for the alternative hypothesis. Our result could actually be written:

> We tested the hypothesis that the effect of density on egg production in lim-
> pets depends on the season in which they are reproducing. We found no
> evidence for an interaction between density and season ($F = 0.0711$; $df = 1$,
> 20; $p = 0.79$), indicating that the effects of density and season are additive.

We could stop the analysis there. We have analysed data from an experiment designed to test the hypothesis, and R has again provided an
answer. However, we probably want more. Perhaps we want to know what
the estimate of egg production at low density is in each season (intercepts;

density-independent egg production). We may also want to know what the
rate of density dependence is, given that it appears to be common.

6.3.7 INTERPRETATION: THE summary() TABLE

We can use the summary table to identify all of this:

```
summary(limp.mod)

##
## Call:
## lm(formula = EGGS ~ DENSITY * SEASON, data = limp)
##
```

```
## Residuals:
##      Min        1Q    Median       3Q       Max
## -0.65468  -0.25021  -0.03318   0.28335   0.57532
##
## Coefficients:
##                          Estimate Std. Error t value Pr(>|t|)
## (Intercept)              2.664166   0.234118  11.380 3.45e-10
## DENSITY                 -0.033650   0.008259  -4.074 0.000591
## SEASONsummer            -0.812282   0.331092  -2.453 0.023450
## DENSITY:SEASONsummer     0.003114   0.011680   0.267 0.792533
##
## (Intercept)             ***
## DENSITY                 ***
## SEASONsummer            *
## DENSITY:SEASONsummer
## ---
## Signif. codes:
## 0 '***' 0.001 '**' 0.01 '*' 0.05 '.' 0.1 ' ' 1
##
## Residual standard error: 0.4079 on 20 degrees of freedom
## Multiple R-squared:  0.7135, Adjusted R-squared:  0.6705
## F-statistic:  16.6 on 3 and 20 DF,  p-value: 1.186e-05
```

The output from **summary**() is more manageable here than it was with the two-way ANOVA. We can see that it has four sections. First, it provides a restatement of your model by repeating your usage of **lm**()—the call. Next it provides the range, interquartile range, and median of the residuals as a quick, additional check of the residuals. These should not be relied upon as a sole diagnostic and should be used with the more substantial plotting methods described above.

The next section of the output is the coefficient table. This is one of the most interesting tables a statistics package can produce associated with an ANCOVA, as it provides, one way or another, insight into the estimates for the lines that you have fitted—these are the coefficients that quantify the important biological relationships. If you have used other packages, you will have seen similar tables. Here is how to interpret the table in R.

First, don't forget that R works alphanumerically. In this current example we should thus expect to see, for example, the season 'spring' reported before the season 'summer', because spring comes before summer. We do like saying that.

Let's start our interpretation from the bottom, where we find some generic statistics and a commonly used estimate of the variance explained by the model, R^2. The model we have fitted explains 67% of the variation in egg production, and has a significant fit to the data ($F = 16.6$, $df = 3$, 20, $p < 0.001$), leaving a residual standard error of 0.4079 with 20 degrees of freedom. Hopefully, you were expecting 20 degrees of freedom for error: 24 data points and 4 estimated coefficients give 20 – 4 error degrees of freedom.

Now for the fun part. First, recall that R behaves alphanumerically. Spring comes before summer. We told you we like saying that. Second, note that R has labelled aspects of rows three and four with the word 'summer'—this should give something away.

Looking closely at the top two rows, we can see two words in the leftmost column under the word 'Coefficients': Intercept and Density. The equation for the line we are fitting is EGGS = Intercept (b) + slope (m) × DENSITY. This should be reassuring. R has told us that here is the estimate of an intercept and a slope. And, we know that it is the intercept and slope estimate for the spring data. Spring comes before summer.

Specifically, the model (equation for a line) for egg production in spring is:

$$\text{EGGS}_{\text{spring}} = 2.66 - 0.033 \times \text{DENSITY}.$$

Looking at rows three and four, we think you may now know what we are supposed to do. You should recall from our previous discussions of treatment contrasts that these will be differences. Specifically, the third line, labelled SEASONsummer, is very specifically the difference between the spring and the summer intercept. More biologically, it is the change in egg production that arises from shifting the season from spring to summer. The value is –0.812 eggs.

If this is the difference between the spring and summer intercepts, then adding this number of eggs to the estimate of the intercept for spring should provide our estimate of the intercept for summer.

Likewise, perhaps even logically because of the words 'summer' and 'DENSITY', the fourth line is the difference between the slopes for spring and summer. More biologically, it is the change in the rate of density dependence that arises from shifting from spring to summer. The value is 0.003. If this is the difference in slopes, then adding this number to our estimate of the slope for spring should provide our estimate for the summer slope. Here is the maths:

$$EGGS_{spring} = 2.66 - 0.033 \times DENSITY,$$
$$EGGS_{summer} = (2.66 - 0.812) + (-0.033 + 0.003) \times DENSITY,$$
$$EGGS_{summer} = 1.84 - 0.03 \times DENSITY.$$

We also highlight the t-values and p-values in this table. As we covered in Chapter 5, a t-test essentially evaluates whether two values are different by asking if the difference between the two values differs from zero. Here, the treatment contrasts, comparing for example the difference in the intercepts for each season, makes sense. The t-value and p-value reveal whether the season intercepts and slopes differ. If the t-value is small and the p-value large, we are unable to reject the null hypothesis that the two values are the same (there is no difference). We can reject this null hypothesis for intercepts, but not slopes.

So, via just the summary table, we can draw the following conclusions. Summer reduces egg production compared with spring, on average, by 0.812 eggs. In summer, the rate of decline in egg production with density was slightly less than in spring (+0.003 eggs/density) but we could not conclude that this change was different from 0. Finally, as a result of this last piece of information, there is no evidence that the effect of density depends on season. Good, the results of the summary table and ANOVA agree, as they should!

Having a priori defined the hypothesis we were testing as 'the effect of density on egg production depends on season', with an associated null hypothesis that it does not, we find that we cannot reject the null hypothesis.

There is no evidence in these data that the effect of density on egg production depends on season. You never thought you'd know this much about
limpets.

6.3.8 PUTTING THE LINES ONTO THE FIGURE

Our final instalment of the linear model centres on a robust method
for producing a figure that combines the raw data with model fits—that
is, the lines estimated by the ANCOVA. The method builds on your
understanding that R provides you, in the summary table, with the coefficients for two lines. In contrast to our approach for the simple linear
regression (Chapter 5), we present here a very generic method, useful
for models of arbitrary complexity (i.e. more than two lines, non-linear
models, etc.).

We have used R to fit a model, and the object has stored in it the coefficients for the lines. If you want to see them, use **coef**(limp.mod). In order
to add lines to a graph, we need to create (predict) some 'y' values from a
sensible range of 'x' values, with the coefficients defining the new y values.
We have to take, for example,

$$EGGS_{Spring} = 2.66 - 0.033 \times DENSITY$$

and provide some reasonable estimates of DENSITY. You've probably done
something like this in Excel. It is worth bearing in mind now that if we
'just' wanted the lines, all we would need would be two points. But if we
want to be flash and fancy, as is expected these days, and add transparent
bands representing the 95% confidence interval around the fitted values
(doesn't that sound flash?), we need more than two points, because the
standard errors around a fitted line are decidedly non-linear. We'll see that
below. To start this process, we invoke two functions: **expand.grid**() and

predict().

expand.grid() is a function that generates a grid of numbers—
essentially it builds a factorial representation of any variables you provide
to it, and returns a data frame. For example:

```
expand.grid(FIRST = c("A", "B"), SECOND = c(1, 2))

##    FIRST SECOND
## 1      A      1
## 2      B      1
## 3      A      2
## 4      B      2
```

Here, the first and second columns are each of the variables and the grid is expanded like a factorial design.

predict() is a function that generates fitted values from a model. It requires at least one argument, but works most effectively for us with three: a model, a set of new '*x*' (explanatory) values at which we want to know '*y*' values, generated according to our model, and another argument, interval, which provides a rapid way to get the 95% confidence interval, doing some hard work for us.

If we use **predict()** with only a model object as an argument, it returns to us the predicted value of '*y*' at each of the '*x*'s in our original data frame. For our limp.mod model and the original data frame of 2 seasons × 4 density treatments × 3 replicates, we should see 24 predicted values returned, with sets of 3 (the replicates) being identical:

```
predict(limp.mod)

##         1         2         3         4         5         6
## 2.3949692 2.3949692 2.3949692 1.6075953 1.6075953 1.6075953
##         7         8         9        10        11        12
## 2.1594217 2.1594217 2.1594217 1.3938428 1.3938428 1.3938428
##        13        14        15        16        17        18
## 1.6546769 1.6546769 1.6546769 0.9358016 0.9358016 0.9358016
##        19        20        21        22        23        24
## 1.1499321 1.1499321 1.1499321 0.4777604 0.4777604 0.4777604
```

However, let's assume that we would specifically like a prediction of the number of eggs at a set of specific densities. First, we create the 'new *x*'s' (DENSITY) at which we want values of *y* (EGGS). Here we use **expand.grid()** to generate a set of 'new *x*' values, *and* we take special care to label them as the column name in the original dataset, DENSITY.

We know the density varies from 8 to 45, so let's use the function **seq()** to get ten numbers in between 8 and 45:

```
# make some new DENSITY values at which we request predictions
new.x <- expand.grid(DENSITY =
                     seq(from = 8, to = 45, length.out = 10))

# check it worked
head(new.x)

##      DENSITY
## 1   8.00000
## 2  12.11111
## 3  16.22222
## 4  20.33333
## 5  24.44444
## 6  28.55556
```

Let's add **SEASON** to this grid now. **SEASON** has two levels, spring and summer. We can add these levels to the 'new x's' as follows. Note how we use the function **levels()**, against the dataset, to extract the information we desire:

```
# make some new DENSITY values at which we request predictions
new.x <- expand.grid(
   DENSITY = seq(from = 8, to = 45, length.out = 10),
   SEASON = levels(limp$SEASON))

# check it worked
head(new.x)

##      DENSITY SEASON
## 1   8.00000 spring
## 2  12.11111 spring
## 3  16.22222 spring
## 4  20.33333 spring
## 5  24.44444 spring
## 6  28.55556 spring
```

Notice that **expand.grid()** has now created a factorial representation of DENSITY (four levels) and SEASON (two levels), such that each and every DENSITY:SEASON combination is represented. You may want to just type new.x to see it all. Now, we can embed these 'new x's' into the function **predict()**, in the argument known as ... wait ... newdata!

We will use **predict()** with three arguments: a model, a value for newdata, and a request for confidence intervals. We love this part. And we assign what predict returns to an object called new.y:

```
# generate fits and confidence interval at new.x values.
new.y <- predict(limp.mod, newdata=new.x,
                 interval = 'confidence')

# check it!
head(new.y)

##        fit      lwr      upr
## 1 2.394969 2.019285 2.770654
## 2 2.256632 1.931230 2.582034
## 3 2.118294 1.834274 2.402315
## 4 1.979957 1.724062 2.235852
## 5 1.841619 1.595998 2.087241
## 6 1.703282 1.447918 1.958646
```

This is just awesome. We have new.x, which looks like a version of the explanatory variables. And we have new.y, which is estimated using the coefficients from the model we have actually fitted, along with the 95% confidence interval around each value, with nice names — fit, lwr, and upr. Wowsa!

The next step we call 'housekeeping'. Housekeeping is an important part of using R, and at this point in the plotting cycle we advocate a bit of it. What we suggest is that you combine the new *y*'s with the new *x*'s, so that we have a clear picture of what it is you've made. We can do that with the function **data.frame()**. We call the data frame addThese, 'cause we are gonna add these to the plot.

And we make one *very important change*: we rename fit produced by **predict** to EGGS to match the original data... **rename()** is another cool little *dplyr* function!

```
# housekeeping to bring new.x and new.y together note that we
# rename fit to be EGGS matching the original data
addThese <- data.frame(new.x, new.y)
addThese <- rename(addThese, EGGS = fit)
# check it!
head(addThese)

##     DENSITY SEASON     EGGS      lwr      upr
## 1   8.00000 spring 2.394969 2.019285 2.770654
## 2  12.11111 spring 2.256632 1.931230 2.582034
```

```
## 3 16.22222 spring 2.118294 1.834274 2.402315
## 4 20.33333 spring 1.979957 1.724062 2.235852
## 5 24.44444 spring 1.841619 1.595998 2.087241
## 6 28.55556 spring 1.703282 1.447918 1.958646
```

So, now you have a new, small data frame that contains the grid of seasons and densities, as well as the predicted values and 95% CI at each of these combinations. You did not need to specify the coefficients, or the equations for the lines. **predict()** does all of that for you. You did not need to add and subtract 1.96*standard error to and from each value either to generate the CI... **predict()** does all of that for you. **predict()** is your super-friend.

Before we finish with adding these data to the figure, we want to show how flexible this method is. Let's modify the code so that we get the predicted number of eggs for each season at the mean limpet density:

```
# new.x with DENSITY set to mean
  new.x <-expand.grid(DENSITY =mean(limp$DENSITY),
                      SEASON = levels(limp$SEASON))

# predictions
predEgg <- predict(limp.mod, newdata = new.x)

# housekeeping
EggAtMeanDens <-data.frame(new.x, predEgg)
head(EggAtMeanDens)

##    DENSITY SEASON predEgg
## 1    24.5 spring 1.83975
## 2    24.5 summer 1.10375
```

6.3.9 THE FINAL PICTURE USING **ggplot()**

You are nearly there. You've carried out quite a substantial ANCOVA of an experiment. You've plotted the data, you've made a model, you've checked the assumptions, and you've interpreted the model. You've got some predictions as well, detailing how the model has fitted lines to your data. The final step, the last effort you may want to make, is to add these lines to the figure (e.g., Figure 6.8). An informative figure is worth a thousand words. If you can provide a figure that a reader looks at and, as a result, knows the question *and* the answer to the question, you will have communicated science effectively.

Here is the last bit of code to do this:

```
# raw data plot (you don't need to write this again...)

ggplot(limp, aes(x = DENSITY, y = EGGS, colour = SEASON)) +

  # first add the points
  geom_point(size = 5) +

  # now add the fits and CIs
  # note we don't need to specify DENSITY AND EGG
  # they are inherited from above!
  geom_smooth(data = addThese,
              aes(ymin = lwr, ymax = upr,
                  fill = SEASON), stat = 'identity') +

  #now adjust the colours
  scale_colour_manual(values = c(spring="green", summer="red")) +
  scale_fill_manual(values = c(spring="green", summer="red")) +

  # theme it
  theme_bw()
```

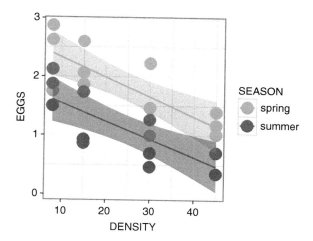

Figure 6.8 Fitted lines for both seasons, added to the raw data for egg production by limpets. These lines we produced using a workflow that involves making predictions using the model via the function `predict()`. This produces lines that do not extend beyond the range of the data, which is good. The workflow to prepare this figure is very general and can be applied to any linear model.

That ggplot code...

There is quite a bit going on in that code, but we've made every attempt to break it up and add annotation. The original raw data-plotting syntax is in there, but we've added the use of **geom_smooth**() to add the fits and CIs. We draw your attention to the annotation that says:

```
# note we don't need to specify DENSITY AND EGG
# they are inherited from above!
```

geom_smooth() is quite smoooooth. So were you. You made sure the names of ALL variables in the addThese data frame matched exactly what are in the original data. So, while we provide **geom_smooth**() with some new aesthetic boundaries for the CIs, it inherits (i.e. assumes) that the 'x =' and 'y =' variables are the same. Woot! So all we needed to do was specify ymin and ymax.

Oh, and stat = 'identity'. We covered that in the χ^2 example. It tells **geom_smooth**() to use what we give it, and not to try and calculate anything fancy-shmancy for us. That's all.

Just to help you along, a concise ANCOVA analysis script is presented in Box 6.1. It probably matches your own if you've been working along! The script is short and provides an archived (save it!), annotated (for you and your colleagues), and cross-platform record of your analysis.

6.4 Overview: an analysis workflow

Hopefully, it is clear that we have introduced you to a workflow for the analysis of your data using R. We have used this workflow consistently for a χ^2 contingency table test, a t-test, a linear regression, a one- and two-way ANOVA, and an ANCOVA. You've been introduced to several tools and tricks and insights into how to fit and interpret R's code and output. We review here nine steps to a happy R-life:

1. Enter your data in a spreadsheet program, check it, and save it as a comma-separated values file.

2. Start R, wipe its brain of all previous data—rm(list=ls())—and get your libraries sorted.

3. Import your data into R and check this has worked, for example using **glimpse()**.

4. Make a picture that reflects your questions and that will help you specify your model to answer your question.

5. Specify a statistical model to capture the hypothesis you are testing.

6. Assess whatever assumptions are important to the type of modelling you have decided to implement (use **autoplot()**).

7. Interpret the model with **anova()** and **summary()**.

8. Add fitted lines and CIs, when appropriate, to the plot; make the figure that is worth 1000 words.

9. Keep your data file and script file very safe (and backed up in a separate location). With these, you can recreate all your graphs and analyses with a few keystrokes in a few minutes at most.

If you adhere to this basic workflow for data analysis, and stick to *Plot -> Model -> Check Assumptions -> Interpret -> Plot Again*, you will have a very solid and efficient foundation for using R. Remember that nothing beats an effective picture for communicating your findings, and that the assumptions are just as important as the predictions.

Box 6.1: The final ANCOVA script

```
# ANCOVA Analysis of limpet reproduction
#libraries
library(dplyr)
library(ggplot2)
library(ggfortify)
#clear the decks
rm(list=ls())
# Read the limpet data and check the structure
limp <- read.csv("limpet.csv")
glimpse(limp) # make the first plot to explore the data
ggplot(limp, aes(x = DENSITY, y = EGGS, colour = SEASON)) +
  geom_point() +
```

┤ **Box 6.1: (continued)** ├

```
scale_color_manual(values = c(spring="green", sum-
mer="red")) +
  theme_bw() # make the ANCOVA model using lm; confirm
what values are returned
limp.mod <- lm(EGGS   DENSITY*SEASON, data = limp)
# check the diagnostic plots for this model
autoplot(limp.mod, smooth.colour=NA)
# use anova() and summary() to interpret the model
anova(limp.mod) # sequential sums of squares table
summary(limp.mod) # coefficients table
# re-make the figure and add fitted lines
# FIRST--new x values--where do we want predicted eggs es-
timated?
new.x <- expand.grid(
  DENSITY = seq(from = 8, to = 45, length.out = 10),
  SEASON = levels(limp$SEASON))
# use predict() and data.frame() to generate the new y's
# collect them (housekeeping) for plotting in object called
preds.for.plot
new.y<- predict(limp.mod, newdata = new.x, interval = 'con-
fidence')
# housekeeping to bring new.x and new.y together
addThese <- data.frame(new.x, new.y)
addThese<-rename(addThese, EGGS = fit)
# remake the figure and add lines and CI
ggplot(limp, aes(x = DENSITY, y = EGGS, colour = SEASON)) +
  geom_point(size = 5) +
  geom_smooth(data = addThese,
          aes(ymin = lwr, ymax = upr, fill = SEASON),
          stat = 'identity') +
scale_colour_manual(values = c(spring="green",
summer="red")) +
scale_fill_manual(values = c(spring="green",
summer="red")) +
theme_bw()
```

7

Getting Started with Generalized Linear Models

7.1 Introduction

The previous two chapters have focused largely on the linear model, that class of model that includes our favourites, including regression, ANOVA, and ANCOVA. The response variables in these tools share some important features in common: we assume they are continuous variables that can take both positive and negative values and can be fractions. We also assume these data are unbounded, though in practice they may not be (e.g. body size can't be negative). These models all also assume normally distributed residuals. And they also assume something special: a constant mean–variance relationship (remember that panel in the diagnostics?).

We always checked these assumptions and found them not badly violated, so it was sensible to use general linear model. Often, however, we work with response variables for which the normality assumption is violated. We often know this in advance, because in the biological sciences we frequently collect data that are bounded and have other features that make this assumption fail. Sometimes we collect data that are integer-valued and bounded, that can not be fractions or negative, like the number of babies.

Getting Started with R Second Edition. Andrew Beckerman, Dylan Childs, & Owen Petchey:
Oxford University Press (2017). © Andrew Beckerman, Dylan Childs, & Owen Petchey.
DOI 10.1093/oso/9780198787839.001.0001

Sometimes we collect data that are strictly binary (e.g. presence/absence data). These types of data can violate the assumptions of the general linear model.

Historically, for these types of data, we often invoked *transformations*. We may have used a `log10` transformation for counts, or an `arc-sin(sqrt())` transformation for proportions. But with the advent of computing speed on our desktop, and developments in statistics, we can now just use a powerful and effective tool: the generalized linear model (GLM). However, this has not stopped people from discussing trans-formations at length. Just search the journal *Methods in Ecology and Evolution* for the three papers since 2010, or look on their Facebook (yes) page for the 'should you log-transform counts' saga (spoiler—sometimes it is OK).

7.1.1 COUNTS AND PROPORTIONS AND THE GLM

Before we dive into using a GLM, let's reveal a bit more about bio-logical data that requires the GLM tool. Count data come in many forms. Biological questions often involve counting a number of things in a unit of time or space, such as the number of individuals in a population, species in a patch, or parasites present within an individual. We usually want to relate these counts to other variables. For example, we might be interested in how the number of offspring produced by a female over her lifetime is related to body mass or age. These are questions where the response variable is counts and the goal is to understand how the rate of occurrence of events (e.g. births) depends on the other variables (e.g. body size). Count data are bounded between zero and infinity, violate the normality assumption, and don't have a constant mean–variance relationship.

Data relating to proportions are also common. A common type of data captures whether or not an event happens, such as does an animal die, does a plant flower, or is a species present in a grid cell? Or we may collect data on sex ratios. Once again, we usually need to relate these occurrences to one or more explanatory variables. For example, we might be interested in whether death is related to the concentration of pesticide an insect is exposed to. This is a question where the response variable is either *binary* (i.e. an indicator variable of dead or alive) or another kind of *count*. In the case of survival, we can work either with an indicator variable that describes whether each individual survives (binary, 0 or 1), or with a summary count that describes how many individuals in a group have died. In both cases the goal is the same: we want to understand how the *probability* of an event occurring depends on the explanatory variables. These data are called binomial. They are also bounded, violate the normality assumption, and don't have a constant mean–variance relationship.

We think you probably get it . . . boundedness and non-constant mean–variance relationships lead to the *Generalized Linear Model*. Here we introduce the GLM as a solution to these problems, an alternative statistical modelling framework that properly accounts for the properties of such variables.

7.1.2 KEY TERMS FOR GLM MODELS

In the following sections, we show the basics of how to work with GLMs in R. This is not an easy task, as the theory and implementation of a GLM are quite different from a linear model. However, we aim to explain, in practical terms and in our 'getting started' voice, some of the nuances, details, and terminology associated with GLMs.

We start by introducing three important terms and their meaning. We will use these in the subsequent example, and want to provide a semi-formal definition in advance. And one of them is the 'family'. Who can argue with learning about family?

1. *Family* This is the probability distribution that is assumed to describe the response variable (also referred to as the *error structure*). By the way, a probability distribution is just a mathematical statement of how likely different events are. The Poisson and binomial distributions are examples of families.

2. *Linear predictor* Just as in a linear model, there is a linear predictor, i.e. an equation that describes how the different predictor variables

(explanatory variables) affect the expected value of the response variable (e.g. birth rates or probabilities of death in the discussion above).

3. *Link function* This one can be tricky at first glance, but, essentially, the clue is in the name. The link function describes the mathematical relationship between the expected value of the response variable and the linear predictor—it 'links' these two parts of a GLM.

Don't worry if these definitions aren't totally clear at the moment. Return to them after you've been through this chapter, and they should (will, we promise) make more sense.

We're only going to look at one type of GLM, the one that often is a good starting point when you have count data. We will also always say 'general linear model' to refer to models fitted with **lm**() and 'generalized linear model' for those fitted with **glm**(). Finally, there is a lot more 'theory' in this chapter because experience has shown us that even when people have been taught GLMs, they miss important details like the significance of link functions... we think you'll appreciate this bit of theory, and it also helps you understand better the theory for the general linear models covered in the previous chapters.

7.2 Counts and rates—Poisson GLMs

7.2.1 COUNTING SHEEP—THE DATA AND QUESTION

We start by counting sheep—don't go to sleep. This is an example where the response variable is a count variable like we discussed above. Key to understanding the question these data were collected to answer is to remember that, with such counts, our the goal is to understand how the *rate* of occurrence of events (counts of babies produced) depends on one or more explanatory variables. That might seem a little cryptic at the moment, but trust us, it will make sense by the end of this section.

Our example can be thought of as a study of natural selection. Hirta is a small island off the west coast of Scotland that is home to an unmanaged population of feral Soay sheep (unmanaged *and* feral; yikes!). These sheep have been the subject of a great deal of ecological and evolutionary research, including studies of factors related to female (i.e. ewe) fitness. One way to measure fitness is to count the total number of offspring born to a female during her life, called 'lifetime reproductive success'. So, our response variable is counts of offspring. Lifetime reproductive success data are a good example of counts that are often poorly approximated by a normal distribution. It seldom appropriate to model them with a general linear model.

Let's assume we have measured the lifetime reproductive success of some ewes in the study population, alongside a standardized measure of average body mass (kg). The question we focus you on is whether lifetime reproductive success increases with ewe body mass; do bigger mums make more babies? If it does, and there are heritable differences in mass (offspring are like their parents), then in the absence of any biological constraints (a very strong assumption!) we should expect selection to increase body mass over time. Our goal here is to evaluate the hypothesis that body mass and fitness are positively associated in Soay sheep.

The data for this example are stored in a file called soaysheepfit-ness.csv (get it the same way you did for all other datasets). First we need to read this data into R and check its structure. Make sure you set this script up exactly the same way you did before, clearing the decks and getting **ggplot2** and **dplyr** loaded and ready to use. Then:

```
soay <- read.csv("soaysheepfitness.csv")
glimpse(soay)
```

The data are very simple—there are 50 observations and only two variables. The body.size variable is the standardized measure of average body mass (in kg) of each ewe, and fitness is her lifetime reproductive success.

As always, we should make a graph to summarize the key question. A scatterplot of fitness against body mass is the obvious choice (Figure 7.1). In addition to plotting the raw data, we can also display what a linear model might look like fitted to the data, and a non-linear relationship

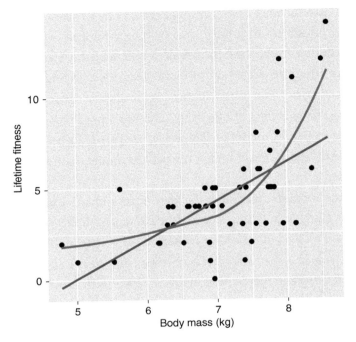

Figure 7.1   Fitness (number of offspring) versus mother's weight.

too. We can actually do both of these super-quickly using **geom_smooth**() from *ggplot2*:

```
ggplot(soay, aes(x = body.size, y = fitness)) +
  geom_point() +
  geom_smooth(method = "lm", se = FALSE) +
  geom_smooth(span = 1, colour = "red", se = FALSE) +
  xlab("Body mass (kg)") + ylab("Lifetime fitness")
```

The blue line produced by **geom_smooth**(method = "lm", se = FALSE) shows a fitted linear regression line (remember using this in Chapter 4?), while the red line generated by **geom_smooth**(span = 1, colour = "red", se = FALSE) shows the non-linearity using a more flexible statistical model. The curve is produced by something called a local regression—we're not going to worry about how this works here other than to point out that the span = 1 bit controls how 'wiggly' the line is (change it to 0.5 if you don't believe us).

Both lines clearly indicate a strong positive relationship between fitness and body size, whereby larger ewes have more offspring over their lifetime. This probably isn't very surprising. If it is, back to school. Big mums simply have more resources to 'spend' on reproduction. But it is also fairly obvious that the straight-line relationship (blue line) is rubbish and doesn't capture the general pattern very well. There is a degree of upward curvature in the fitness–size relationship, captured by the wiggly red line.

Is this a problem? Probably. We might deal with this non-linearity by using some kind of transformation or including a squared term in our regression model. However, there are other, more subtle problems in these data. To truly understand these problems, and the ultimate value of the generalized linear model, we are going to first carry out the 'wrong' analysis using the familiar linear regression model. We'll then use our favourite diagnostic plots to identify these problems. This is a much better strategy than trying to identify them by endlessly staring at the raw data. Once we understand the data and their problems, we'll carry out the 'right' analysis using generalized linear model. This approach will help you understand why GLMs are useful, and on the way you'll learn a little bit about how they work.

7.3 Doing it wrong

You know how to fit a general linear model with the **lm**() function and generate diagnostic plots (Figure 7.2) with the **autoplot**() function from *ggfortify* . . .

We hope you did make the model. And these look terrible, don't they? Finally, terrible diagnostics. But let's remember that we have just fitted a straight-line relationship to data that look non-linear and assumed normally distributed residuals with constant variance. The diagnostic plots are telling us there are several problems with these assumptions. In fact, they are telling us that we have violated nearly all of them.

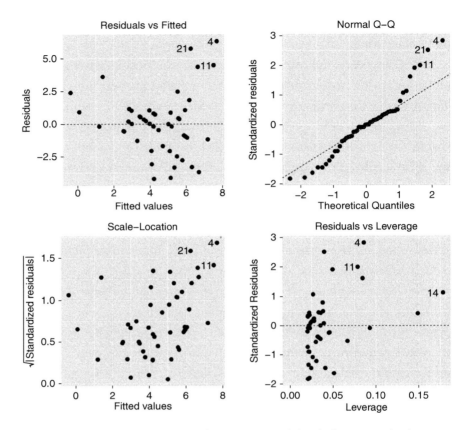

Figure 7.2 Diagnostic graphs for linear model of fitness – body mass relationship.

7.3.1 DOING IT WRONG: DIAGNOSING THE PROBLEMS

The plot of residuals vs fitted values (upper left panel) tells us that the systematic part of the model is not very good, i.e. there is a clear pattern in the relationship. Actually, we can say a bit more than this. The U-shape indicates that a straight-line model fails to account for curvature in the relationship between our two variables. We underestimate fitness at small body sizes, then overestimate it at medium sizes, and then underestimate it at large sizes.

The normal Q–Q plot (upper right panel) is also a problem. Most of the points should lie on the dashed line. Instead, points corresponding to the most positive 'theoretical quantiles' are consistently above the line and those associated with the most negative values are below it. It looks like our normality assumption is way off. Again, we can say more if we know what we're looking for. This particular pattern occurs because the distribution of the residuals is not symmetric; it is skewed to the right. If you want, make a histogram, go ahead . . .

The scale–location plot (bottom left panel) shows a positive relationship between the absolute size of the residuals and the fitted values. This reflects the way the fitness values are scattered around the red line in Figure 7.1. There is more vertical spread in the data at larger values of fitness. We can say there is a positive mean–variance relationship: large predicted fitness values are associated with more variation in the residuals. This is typical of count data such as these.

At least the residuals–leverage plot (bottom right panel) isn't too bad! There are no really extreme standardized residual values, so we don't seem to have any obvious outliers, and none of the observations are having too much of an effect on the model.

Overall, this exercise in fitting what we know to be the wrong model, the normal linear regression model, is not doing a very good job of describing these data. We're going to fix the model, but first we need to learn a little bit about a new distribution.

7.3.2 THE POISSON DISTRIBUTION—A SOLUTION

The problem with our linear regression model is that the normality assumption just isn't appropriate for raw, untransformed count data. Why are count data unlikely to be normally distributed? If we think about the properties of the normal distribution, we can list a few fairly obvious reasons:

1. The normal distribution concerns continuous variables (i.e. those that can take fractional values). However, count data are discrete. It is possible for a ewe to produce 0, 1, 2, or 3 lambs, but impossible for her to give birth to 2.5 offspring.

2. The normal distribution allows negative values, but count data have to be positive (we'll include 0 here). A ewe may have 0 offspring, but she certainly cannot have –2 offspring. Isn't biology cool?

3. The normal distribution is symmetrical, but counts are often (but not always) distributed asymmetrically, in part because they cannot be negative. This might not be obvious to you, but it's true.

The normal distribution is just not a very good model for many kinds of count data. On the other hand, the Poisson distribution *is* a good starting point for analysis of *certain* kinds of count data. This isn't a statistics book, so we aren't going to explain the properties of the Poisson distribution in detail. A visual description is useful, though. Figure 7.3 shows three Poisson distributions, each with a different mean. The x-axis shows a range of different possible values (counts), and the y-axis shows the probability of each value.

We can see why the Poisson distribution is such a good candidate for count data, and for the problems we identified above, when fitting the linear regression model:

- Only discrete counts (0, 1, 2, 3, . . .) are possible. These are bounded at 0, and while very large counts are possible, the probability of one occurring is very low.

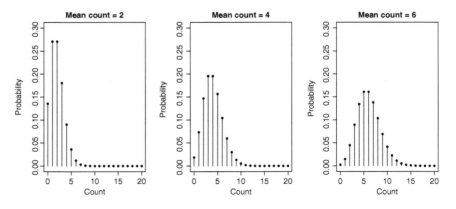

Figure 7.3   Examples of the Poisson distribution

- The variance of the distribution increases as the mean of the distribution is increased. Visually, this corresponds to a widening of the base of the distribution as the mean grows.

The Poisson distribution is best suited for analysing 'unbounded' count data. *Huh? What do you mean by 'unbounded'? You just said the data were bounded!!!* Yes, we did. But this just refers to the fact that there is no upper limit to the values the count variable might take. For example, a female Soay sheep could produce 2, 5, or even 10 offspring over her lifetime. In reality there must be some biological constraints on lifetime reproductive fitness—a ewe will never produce 100 offspring—but that limit isn't something we know. The unbounded count assumption is just a useful approximation to reality.

So how do we model the relationship between the fitness of Soay ewes and their body mass using a Poisson distribution? By using a generalized linear model, of course! However, before we can use a GLM, we have to understand its component parts. Sorry, but we need a tiny bit more theory.

7.4 Doing it right—the Poisson GLM

7.4.1 ANATOMY OF A GLM

We mentioned three important terms related to the GLM at the beginning of this chapter: the 'family', the 'linear predictor', and the 'link function'. You have to understand at least roughly what these mean to use GLMs without making a terrible mistake. We're just going to give you a quick tour so that we can finally get up and running with GLMs in R.

The family

The family (or 'error') part of a GLM is not too difficult to understand. This bit just determines what kind of distribution is used to describe the response variable. A *general* linear model always assumes normality— that's pretty much its defining feature in fact—but with a *generalized* linear model there are several different options available. We can use a Poisson distribution (no surprises there), a binomial distribution (more on that later), or a gamma distribution (for positive-valued, continuous variables), plus a few other more exotic versions. Each of these is appropriate for specific types of data, but taken together, these different GLM families allow us to work with a wide range of response variables. GLMs supercharge your data analysis.

The linear predictor

Although you might not realize it, you already know about the GLM linear predictor. Every time you build a model using **lm**() you have to supply it with at least some data and a special little R formula to define the model. The job of that formula is actually to define the linear predictor (remember, general linear models have these too).

Let's try to understand that linear predictor. It's most easily understood with simple regression. We can use the Soay example for this. We fitted the bad regression using **lm**(fitness $\sim$ body.size, . . .). That tells R to 'build me a model for the predicted fitness, with an intercept and a body size slope term.' It looks like this:

$$\text{Predicted Fitness} = \text{Intercept} + \text{Slope} \times \text{Body Size}.$$

We don't have to tell R to include the intercept; it adds it automatically because models without intercepts are seldom sensible. More, wait, most importantly, the bit of that little equation on the right side of the = is the linear predictor! So the linear predictor is really just 'the model', and all those coefficients shown by **summary**() are just estimates of different intercepts and slopes in the linear predictor. We really meant it when we said you already knew about the linear predictor.

By the way, it's called a linear predictor because we 'add up' the component parts, and every part is either an intercept or a slope term. Every single model in the previous couple of chapters—regression, ANOVA, ANCOVA—can be written down as a linear predictor, although models including factors are a little more tricky (so we won't try to do it).

The link function

So that's the linear predictor sorted. What about the link function? This is the one part of GLMs that tends to confuse people. It confuses people a lot. But start simple. We have some problems, Houston, with our diagnostic plots. They are caused by bounded, integer data where the variance increases with the mean, in this case. The combination of choosing the right family and a link function with it can solve these problems.

A good way to start thinking about the link function is to demonstrate why we need it. Think about the Soay sheep fitness linear regression model. Let's plug some numbers in. What happens if the estimated intercept is −2 and the slope is +1.2? Let's try to predict the average number of offspring produced by a 2 kg ewe: −2 + 1.2 × 2 = 0.4. We're only predicting the average, not an actual number, so that seems OK. What about a 1 kg ewe? Now we get −2 + 1.2 × 1 = −0.8. Negative lambs! Not a very sensible prediction.

While the professional sheepologists might argue that 1 kg is not a very realistic mass for a sheep, really, we would prefer to have a model that cannot make impossible predictions. This is a job for the link function. When using a GLM, instead of trying to model the predicted values of the response variable directly, *we model a mathematical transformation of the prediction*. The function that does this transformation is called the link function.

Confused? Fair enough, that last paragraph might be a little difficult to understand. Let's work with the Soays again to get a better idea of what the link function does. If we had used a Poisson GLM to model the fitness–mass relationship, then the model for predicted fitness would look like this:

$$\text{Log[Predicted Fitness]} = \text{Intercept} + \text{Slope} \times \text{Body Size.}$$

The important point to note here is that the link function in this case is the *natural log*. The link function in a standard Poisson GLM is always the natural log.

Now, instead of the linear predictor describing fitness directly, it relates the (natural) logarithm of predicted fitness to body size. This has to be positive, but its log-transformed value can take any value we like (7.1, −2, 0, etc.). The log link function means we have to do a tiny bit of algebra to get the equation for the predicted fitness:

$$\text{Predicted Fitness} = e^{\text{Intercept}+\text{Slope}\times\text{Body Size}}.$$

So... a Poisson generalized linear model for the Soays actually implies an exponential (i.e. non-linear!) relationship between fitness and body mass. Linear model does not mean linear relationship. Remember that. Incidentally, if you remember the diagnostics for the simple regression model you'll probably realize that this exponential relationship could be a good thing. Just look at Figure 7.1 again too.

In summary, a link function allows us to estimate the parameters of a linear predictor that behaves 'correctly', and it does so by moving us from the response 'scale' to the scale of the linear predictor, in this case the natural log scale, defined by the link function. Phew. Still confused?

We've found that a graphical description of the log link function can help (Figure 7.4). The vertical dashed axis represents the whole real number line, encompassing both positive (top) and negative (bottom). This is where we want to be. However, with count data, we are bounded... stuck above 0. This is where the predictions in a Poisson model have to live, i.e. a predicted value has to be greater than zero in order for it to be valid for count data. But to do effective statistics, we need to be on a scale that is unbounded. This is what the link function does. It puts us in a happy

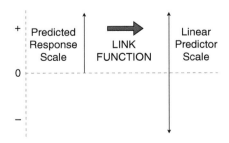

Figure 7.4   Graphical description of the log link function in a GLM.

place, where the linear predictor can live and the response can take any value it likes. The logarithm is the 'right' link function to use here, because it moves us from the positive numbers (predicted average counts) to the whole real number line (the linear predictor).

There are various other terms and ideas we could spend time explaining that are a central part of fitting generalized linear models—likelihoods, deviances, likelihood ratio tests, dispersion, etc.—but we've skimmed through enough already to start working with GLMs. This is supposed to be a practical introduction to GLMs in R, so we'll just briefly explain additional aspects as they crop up.

7.4.2 DOING IT RIGHT—ACTUALLY FITTING THE MODEL

Let's see if we can improve the Soay fitness–mass model by fitting a Poisson GLM. The good news is that you already have most of the skills needed to build this model in R and understand it. It is very easy to do this because you've already mastered **lm**(), **autoplot**(), **anova**(), and **summary**(). We are going to use the same *Plot -> Model -> Check Assumptions-> Interpret -> Plot Again* workflow. Since we've already done the *Plot* part, it's time to get on with the rest.

We don't use **lm**() to build a GLM. Instead we use (drum roll . . .) a function called **glm**(). This works in exactly the same way as **lm**(), though, with the addition that we also have to tell it which family to use:

```
soay.glm <- glm(fitness ~ body.size, data = soay,
                family = poisson)
```

This really is the same as the code for a linear model. We give **glm**() a formula to define the model, some data to work with, and the family to use.

What happened to the link function!? R is very sensible, so if we choose not to specify the link function it will pick a sensible default—the 'canonical link function'. This sounds fancy, but 'canonical' is just another word for 'default'. Remember, this is the log link function for Poisson models. If you want to be explicit about the link function, or change the default, this is easy to do (we'll keep using log). By the way, you can see the canonical links by looking at ?family:

```
soay.glm <- glm(fitness ~ body.size, data = soay,
                family = poisson(link = log))
```

7.4.3 DOING IT RIGHT—THE DIAGNOSTICS

Fitting the GLM was easy. Now, let's look at the diagnostic plots next. These are produced by plotting the model we made, and **autoplot()** can handle a **glm()** just fine. We'll discuss why we use the 'same' diagnostics for the GLM below.

   These plots definitely look better (Figure 7.5):

- The plot of residuals vs fitted values (upper left panel) suggests that the systematic part of the model is now pretty good. There is no clear pattern in the relationship, apart from the very slight upward trend at the end. This is nowhere close to being large enough to worry about— it's driven entirely by only two points on the very right.
- The normal Q–Q plot (upper right panel) is also much better. It isn't perfect, as there is some departure from the dashed line, but we shouldn't expect a perfect plot. Life is never perfect. It is a lot better than the corresponding the plot from the **lm()** model, so it looks like our distributional assumptions are OK.
- The scale–location plot (bottom left panel) seems to show a slight positive relationship between the size of the residuals and the fitted values. If you just focus on the points, there isn't much going on.
- The residuals–leverage plot (bottom right panel) is also fine. There is no evidence that we have to worry about outliers or points having too much of an effect on the model.

Is anything bothering you at this point? Perhaps that 'normal Q–Q plot' seems odd. We fitted a Poisson model, so what's all this talk of normal distributions? Perhaps the scale–location plot seems wrong. Surely if the Poisson model was used we should be hoping for a positive relationship, because the variance should increase with the mean?

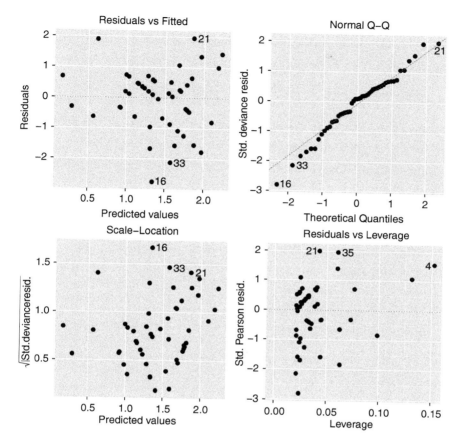

Figure 7.5 Diagnostic graphs for general linear model of fitness–body mass relationship with Poisson error structure and log link function.

Here's the answer(s). When R builds diagnostics for a GLM, it uses something called the standardized deviance residuals. Again, these sound a lot more fancy than they really are. They are a specially transformed version of the raw residuals that make the transformed residuals normally distributed, if (and only if) the GLM family we're using is appropriate.

What this means is, *if the chosen family is a good choice for our data*, then our diagnostics should behave like those from a model with normally distributed errors. You don't need to learn any new skills to evaluate your model with R's GLM diagnostics! The 'normal Q–Q plot' for the transformed residuals *is* checking whether the Poisson distribution is appropriate for the distribution of the residuals, and the scale–location plot *is* checking whether the mean–variance relationship is OK (it will only be patternless if the Poisson distribution is the right model).

7.4.4 DOING IT RIGHT—anova() AND summary()

So far, so good. We have a decent model for the data, so we can finally get back to the original question. Is fitness positively related to body mass? It certainly looks that way, but we still need a *p*-value to convince people. On to step 3 then: testing whether the body mass term is significant.

Once more, you already know how to do this. We use **anova()** (remember, it does not perform an ANOVA). For something produced by **lm()**, **anova()** produces an ANOVA table. Something very similar happens when we use it with a **glm()** model:

```
anova (soay.glm)

## Analysis of Deviance Table
##
## Model: poisson, link: log
##
## Response: fitness
##
## Terms added sequentially (first to last)
##
##
##              Df Deviance Resid. Df Resid. Dev
## NULL                          49        85.081
## body.size   1    37.041       48        48.040
```

The first thing to notice is that it no longer prints an ANOVA table— it shows us an 'Analysis of *Deviance*' table. Don't panic! A lot of this is familiar stuff. The preamble at the top tells you what kind of table it is and reminds you what kind of GLM you used (Poisson, with a log link). It also informs you that this is a sequential table. You know about these. The bit you really care about is that table at the end. We will talk about that in a moment, but first, what is this deviance stuff?

That word 'deviance' is a lot less interesting than it sounds. It is closely related to something called the likelihood, a very general tool for doing statistics. This really isn't the place to explain likelihood, but we'll give you a super-short explanation. Ready? The likelihood of a statistical model, and some data, provides us with a measure of how probable the data would be if they really had been produced by that model. If you know what you're doing, you can use this to find a set of best-fitting model coefficients by picking values that maximize this likelihood thing. Sums of squares, and mean squares allow you to compare different models assuming normality. The likelihood (and deviance) do the same thing for GLMs (and many other kinds of models). OK, that might be the world's shortest introduction to likelihood, but it's good enough for us.

The second thing to notice about the table is that instead of list-ing sums of squares, mean squares, degrees of freedom, F-values, and p-values, we only see degrees of freedom, deviances, and residual de-viances. Arggghhhh, no p-values! The table does tell us that the total deviance in the data (fitness) is 85.081 units and the deviance explained by body size is 37.041 units. That means that nearly half the deviance is accounted for by body size, but what does it mean? Where. Is. Our. p-value!?

There's no p-value here because R wants you to specify what test to use to calculate it. With a GLM, you can choose different types of tests. Here's what you need to do. p-values in the typical GLM involve the χ^2 distribu-tion rather than the F-distribution. Note this does not mean we are doing a χ^2 test:

```
anova(soay.glm, test = "Chisq")

## Analysis of Deviance Table
##
## Model: poisson, link: log
##
## Response: fitness
##
## Terms added sequentially (first to last)
##
##
##              Df Deviance Resid. Df Resid. Dev  Pr(>Chi)
## NULL                          49      85.081
## body.size  1    37.041        48      48.040 1.157e-09 ***
## ---
## Signif. codes:
## 0 '***' 0.001 '**' 0.01 '*' 0.05 '.' 0.1 ' ' 1
```

With this extra detail in place, we can see that that the test statistic is a χ^2 value of 37.041, with one degree of freedom. The *p*-value is very small. This is all rather unsurprising given the strong relationship we can see in the data (Figure 7.1). It is good to be sure, though. Where did that *p*-value come from? It's all about that likelihood again. So, congratulations, you just carried out a likelihood ratio test. That's what **anova**(..., test ="Chisq") is telling R to do. This is highly significant because the deviance associated with the body mass term is so big (trust us on this one). In fact, while you do have to trust us, you can also see that the deviance has dropped from 85 without the body mass explanatory variable to 48 ... lots of variation has been 'explained'. Does fitness vary positively with ewe body size? Yep ($\chi^2 = 37.04$, *df* = 1, *p* < 0.001)! That's what you might report!

We could actually stop the analysis there. We have analysed our data to test a hypothesis about selection on body size. It looks like there is positive selection operating, so maybe one day Soay sheep will be the size of elephants (or maybe not).

However, we may want to understand what our model is telling us beyond 'there is a significant effect of body mass on fitness'. Let's not forget about the summary table of coefficients. You'll be pleased to find out that we can use **summary**() to print this:

```
summary(soay.glm)

##
## Call:
## glm(formula = fitness ~ body.size, family = poisson(link = log),
##     data = soay)
##
## Deviance Residuals:
##      Min        1Q    Median        3Q       Max
## -2.7634   -0.6275    0.1142    0.5370    1.9578
##
## Coefficients:
##              Estimate Std. Error z value Pr(>|z|)
## (Intercept) -2.42203    0.69432   -3.488 0.000486 ***
## body.size    0.54087    0.09316    5.806 6.41e-09 ***
## ---
## Signif. codes:
## 0 '***' 0.001 '**' 0.01 '*' 0.05 '.' 0.1 ' ' 1
```

```
##
## (Dispersion parameter for poisson family taken to be 1)
##
##     Null deviance: 85.081  on 49  degrees of freedom
## Residual deviance: 48.040  on 48  degrees of freedom
## AIC: 210.85
##
## Number of Fisher Scoring iterations: 4
```

This looks familiar . . .

- The first chunk of information at the top is a reminder of the model we are looking at, including information about the family.
- We then see a fairly useless summary of the residuals. Remember, these are those special scaled residuals (the deviance residuals), so a very large number indicates that we have an outlier problem.
- Next come the coefficients. Our model is as simple as they get (it is a line), which means there are only two coefficients: an intercept and a slope. Each estimate has a standard error to tell us how precise it is, a z-value to help us see if the estimate is significantly different from 0, and the associated p-value.
- We are then told about something called the dispersion parameter. This is potentially very important, but now is not the time to explain it. More about this later in this chapter.
- After the dispersion parameter we see summaries of the null deviance, the residual deviance, and their degrees of freedom. The null deviance is a bit like a measure of all the 'variation' in the data. The residual deviance is a measure of what is left over after fitting the model. Big difference => more variation explained.
- Towards the bottom we see the AIC (Akaike information criterion) for the model. We don't use the AIC in this book, but if you are an AIC fan, there's your AIC. Go ahead and use it, for good or for evil.
- You don't need to worry about the 'number of Fisher Scoring iterations'. This is basically a measure of how challenging it was for R to find the best-fit model using all the fancy likelihood stuff.

What do those coefficients tell us? Take a moment to digest what you are looking at; look at the coefficients and look at Figure 7.1. When we teach this stuff, we wait for the inevitable penny to drop . . . the intercept . . . it's negative. But a quick look at Figure 7.1 suggests this isn't what we were expecting . . . . So let's ask again . . . what do these coefficients tell us? Well, they *do not* tell us that a 5 kg female will give birth to an average 0.28 lambs over her lifetime (−2.422 + 0.541 × 5). That should be obvious from the figure we made.

What's going on, we ask again? Let's not even once forget about the link function—the GLM is predicting the natural logarithm of lifetime reproductive fitness, *not* the actual fitness. If we want to know how many lambs a 5 kg ewe is predicted to produce, we must account for what the link function did: lifetime fitness = $e^{(-2.422+0.541\times5)}$ = 1.33 lambs. Phew. Double phew.

There is one more thing we really should check in the summary output. It has to do with that something we called overdispersion. However, we've got a good rhythm going now, so let's come back this later on. The next part is the fun part—making the publication-ready figure that will make us famous for showing Soay sheep will one day be as large as elephants.

7.4.5 MAKING A BEAUTIFUL GRAPH

The method we'll use to overlay the fitted line on the raw data relies substantially on the generic approach to estimating the line that we outlined in the previous chapter, in the ANCOVA example. You just need to learn two new tricks to make it work for a GLM. As before, we use **expand.grid()** to generate a set of 'new *x*' values, remembering to label the single column name as in the original dataset, i.e. body.size (this is not optional—always do this). Rather than trying to guess where to start and end the 'new *x*' values we'll let the min and max functions do the hard work for us, using the **seq()** function to get 1000 numbers in between the minimum and maximum body mass:

```
# note our use of the $ to get the body.size column
min.size <- min(soay$body.size)
max.size <- max(soay$body.size)

# make the new.x values; we use the 'body.size' variable
# name to name the column
# just as it is in the original data.
new.x <- expand.grid(body.size =
                        seq(min.size, max.size, length=1000))
```

Just as with a general linear model, we can now use these 'new *x*'s' with the **predict()** function. Actually, we use **predict()** with three arguments: the GLM model, the value for newdata, and a request for *standard errors*. It is a bit annoying, but **predict()** for **glm()** doesn't have an interval = confidence argument and thus does not return a data frame. You may recall that we asked for interval = confidence in Chapter 6 (in relation to ANCOVA), and that the confidence interval is $\bar{x} \pm 1.96 \times$ SE. We can't get that information automatically. That's why we formally request the standard errors and why we also have to convert the output into a data frame using **data.frame()**:

```
# generate fits and standard errors at new.x values.
new.y = predict(soay.glm, newdata = new.x, se.fit = TRUE)
new.y = data.frame(new.y)
# check it!
head(new.y)

##              fit    se.fit residual.scale
## 1 0.1661991 0.2541777              1
## 2 0.1682619 0.2538348              1
## 3 0.1703247 0.2534919              1
## 4 0.1723874 0.2531491              1
## 5 0.1744502 0.2528063              1
## 6 0.1765130 0.2524635              1
```

The next step is 'housekeeping'. We need to combine the new *x*'s with the new *y*'s so that everything is in one place. We do that using the **data.frame()** function again. We call the data frame to addThese because—just like before—we are going to add these to the plot. And, finally, we mustn't forget one extra *very important change*: we must rename the 'fit' produced by 'predict' to 'fitness' to match the original data. The **rename()** function to the rescue again:

```
# housekeeping to bring new.x and new.y together
addThese <- data.frame(new.x, new.y)
addThese <- rename(addThese, fitness = fit)

# check it!
head(addThese)

##    body.size    fitness     se.fit residual.scale
## 1  4.785300 0.1661991 0.2541777              1
## 2  4.789114 0.1682619 0.2538348              1
## 3  4.792928 0.1703247 0.2534919              1
## 4  4.796741 0.1723874 0.2531491              1
## 5  4.800555 0.1744502 0.2528063              1
## 6  4.804369 0.1765130 0.2524635              1
```

Beauty! Now we have a new data frame that contains the grid of body sizes, as well as the predicted fitness values and standard errors at each of these predictions. We are *nearly* ready to plot everything. But how are we going to add those confidence intervals? One more bit of housekeeping will give us what we need. We need to use the **mutate()** function to simultaneously calculate the CIs and include them in addThese. This last step effectively generates the data frame that interval = confidence did when we used **predict()** with **lm()**:

```
addThese <- mutate(addThese,
                   lwr = fitness - 1.96 * se.fit,
                   upr = fitness + 1.96 * se.fit)
```

OK, now we seem to have everything we need to make our final figure. Let's use the same code format as we used in Chapter 6, and see how it works. But don't study the R code too carefully yet (hint: we might have one more little problem to deal with):

```
ggplot(soay, aes(x = body.size, y = fitness)) +
  # first show the raw data
  geom_point(size = 3, alpha = 0.5) +
  # now add the fits and CIs -- we don't need to specify body.size
  # and fitness as they are inherited from above
  geom_smooth(data = addThese,
              aes(ymin = lwr, ymax = upr), stat = 'identity') +
  # theme it
  theme_bw()
```

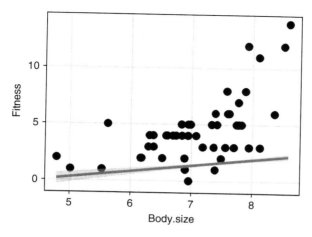

Figure 7.6   Not quite what we wanted... the line is clearly not in the right place.

Uh-oh, that did not work as expected (Fig. 7.6). Something is not right—we forgot about the pesky but vital link function again. The default is for **predict**() to produce predictions on 'the scale of the *link function*' (i.e. on the same scale as the linear predictor). Because we have a log link, this means the predictions are the log of the expected fitness. We wanted predictions of the actual fitness. We've shown you how to get it wrong because you will do this at some point. It helps to diagnose the problem if you've seen it before. It helps us daily.

This mistake is easy to fix. All we have to do is apply the inverse of the logarithm to any *y*-axis variables (i.e. 'unlog' them) in addThese. This is the exponential function. We could do this on the fly inside ggplot, but it is less error-prone to modify everything during the process of making addThese. Make sure you understand what we are doing here. This is what the script/code should look like . . .

```
# range of body sizes
min.size <- min(soay$body.size)
max.size <- max(soay$body.size)

# make the new.x values;
# we use the 'body.size' name to name the column
# just as it is in the original data.
```

```
new.x <- expand.grid(body.size =
                        seq(min.size, max.size, length=1000))

# generate fits and standard errors at new.x values.
new.y = predict(soay.glm, newdata=new.x, se.fit=TRUE)
new.y = data.frame(new.y)

# check it!
head(new.y)

##          fit    se.fit residual.scale
## 1 0.1661991 0.2541777              1
## 2 0.1682619 0.2538348              1
## 3 0.1703247 0.2534919              1
## 4 0.1723874 0.2531491              1
## 5 0.1744502 0.2528063              1
## 6 0.1765130 0.2524635              1


# housekeeping to bring new.x and new.y together
addThese <- data.frame(new.x, new.y)

# check it!
head(addThese)

##   body.size       fit    se.fit residual.scale
## 1  4.785300 0.1661991 0.2541777              1
## 2  4.789114 0.1682619 0.2538348              1
## 3  4.792928 0.1703247 0.2534919              1
## 4  4.796741 0.1723874 0.2531491              1
## 5  4.800555 0.1744502 0.2528063              1
## 6  4.804369 0.1765130 0.2524635              1


# exponentiate the fitness and CI's to get back the
# 'response' scale
# note we don't need rename() because mutate() works with
# the fit values each time, and we 'rename' inside mutate()
addThese <- mutate(addThese,
                   fitness = exp(fit),
                   lwr = exp(fit - 1.96 * se.fit),
                   upr = exp(fit + 1.96 * se.fit))
 # check it!
head(addThese)

##   body.size       fit    se.fit residual.scale  fitness
## 1  4.785300 0.1661991 0.2541777              1 1.180808
## 2  4.789114 0.1682619 0.2538348              1 1.183246
## 3  4.792928 0.1703247 0.2534919              1 1.185690
## 4  4.796741 0.1723874 0.2531491              1 1.188138
## 5  4.800555 0.1744502 0.2528063              1 1.190591
## 6  4.804369 0.1765130 0.2524635              1 1.193050
##        lwr       upr
```

```
## 1 0.7174951 1.943300
## 2 0.7194600 1.946004
## 3 0.7214303 1.948712
## 4 0.7234059 1.951425
## 5 0.7253869 1.954141
## 6 0.7273732 1.956861
```

```
# now the plot on the correct scale
ggplot(soay, aes(x = body.size, y = fitness)) +
  # first show the raw data
  geom_point(size = 3, alpha = 0.5) +
  # now add the fits and CIs -- we don't need to specify
  # body.size and fitness as they are inherited from above
  geom_smooth(data = addThese,
              aes(ymin = lwr, ymax = upr), stat = 'identity') +
  # theme it
  theme_bw()
```

This time our **ggplot()** efforts (Figure 7.7) work as hoped, because everything is on the right scale.

There seems to be a lot going on in that code, but you've seen it all before. The usual raw data-plotting syntax is in there (**geom_point()**), along with **geom_smooth()** to add the fits and CIs. Just as with the ANCOVA example, we have to provide **geom_smooth()** with some new aesthetics for the CIs, but it inherits the 'x =' and 'y =' aesthetics.

Good work. It really isn't all that much effort to produce a sophisticated summary figure from GLM-worthy data in R. It is mostly like plotting

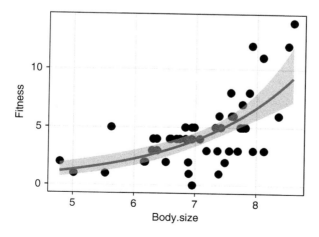

Figure 7.7   The line is now in the correct place. Great!

the results of a linear regression or ANCOVA—just be careful with your housekeeping—as long as you remember the added twist introduced by the link function and the need to 'back transform' from the scale of the linear predictor.

7.5 When a Poisson GLM isn't good for counts

You've now seen all the pieces of the GLM puzzle. We've talked about the basic theory, how to build a GLM in R, interpreting diagnostic plots, making sense of the output from **anova**() and **summary**(), and, finally, how to construct the ever-important figure for publication. We made it look pretty easy too. We did all of this using the simplest possible Poisson regression, and we used data that were very well behaved. It was almost like they were too good to be real data. . . What happens when things are not too good to be true . . .

7.5.1 OVERDISPERSION

It would be very irresponsible to teach you about GLMs in R without discussing the dreaded 'overdispersion'. Overdispersion is just a fancy stats word for 'extra variation'. We're going to explain what this means in the real world, describe some common causes of overdispersion, and then tell you why these matter. We'll finish up by showing you how to detect overdispersion and suggest a couple of possible quick fixes to help you deal with it.

Some GLMs make very strong assumptions about the nature or variability in your data. The Poisson GLM is one such example, as is the binomial GLM. Earlier we noticed that the variance of the Poisson distribution increases with the mean. Actually, the variance *is exactly equal to* the mean for a Poisson distribution. This assumption is only likely be true of real Poisson count data if we are able to include every single source of variation in our analysis.

This is a big assumption, particularly in biology. When you go outside and measure the various features of individuals, populations, or whatever else it is you're studying, no matter how hard you work, there will always be important aspects of your 'things' you can't measure. When these influence your count variable, they create variation you can't account for with a standard Poisson GLM. Even under carefully controlled lab conditions there will be sources of variation you cannot control. Maybe you put all your insects into a controlled environment, but did you ensure they all had the same body size and experienced exactly the same rearing environment? (If you did, well done; that was a lot of work.)

'Ignored stuff' is the root source of the overdispersion problem, but does it arise only from excluded variables? Sadly not. It arises when there is *non-independence* in the data. What does this mean? It means some of your 'things' in your data—individuals, populations, etc.—are more similar to one another than they are to other 'things'. For example, in that insect experiment, individuals from the same clutch/brood/family will be more similar to one another than to a randomly chosen insect, because they share genes and a common rearing environment. We could go on and on about non-independence, but you get the idea.

You might be thinking, 'So what? All models are wrong...' The problem with overdispersion is that if it's there, and you ignore it, it messes up the precious *p*-values. And it does it the worst possible way—they will be anti-conservative. Arghhhhh! Wait, what does that mean? This is a very serious problem. It means you end up with more false positives than you should when the stats are 'working' properly, and *that* means you'll think you've found evidence for things happening that are not in fact happening.

OK, overdispersion is probably common and it's very naughty to ignore it because it breaks *p*-values. What can you do about it? Your first job is to detect it. The good news is that this is reasonably easy to do. We just need to know about a new diagnostic trick.

Look at the summary of the Soay sheep Poisson regression again:

```
summary(soay.glm)

## 
## Call:
```

```
## glm(formula = fitness ~ body.size, family = poisson(link = log),
##     data = soay)
##
## Deviance Residuals:
##     Min        1Q    Median        3Q       Max
## -2.7634   -0.6275    0.1142    0.5370    1.9578
##
## Coefficients:
##              Estimate Std. Error z value Pr(>|z|)
## (Intercept)  -2.42203    0.69432  -3.488 0.000486 ***
## body.size     0.54087    0.09316   5.806 6.41e-09 ***
## ---
## Signif. codes:
## 0 '***' 0.001 '**' 0.01 '*' 0.05 '.' 0.1 ' ' 1
##
## (Dispersion parameter for poisson family taken to be 1)
##
##     Null deviance: 85.081  on 49  degrees of freedom
## Residual deviance: 48.040  on 48  degrees of freedom
## AIC: 210.85
##
## Number of Fisher Scoring iterations: 4
```

The bit we care about here is near the bottom: the 'Residual deviance' (48.040) and its degrees of freedom (48). If a GLM is working perfectly and there is no overdispersion, these two numbers should be equal. The GLM output even declares this constraint, stating 'Dispersion parameter for poisson family taken to be 1'. We can calculate a kind of 'dispersion index' by dividing the residual deviance by the residual degrees of freedom (don't mix up the order of division). This should be about 1—if it is much bigger, the data are overdispersed, and if it is much less, they are underdispersed (which is rare). For our Soay model, there is nothing to worry about as the residual deviance and its degrees of freedom are almost identical.

What if the dispersion index had been 1.2, 1.5, 2.0, or even 10? When should you start to worry about overdispersion? That is a tricky question to answer. One common rule of thumb is that when the index is greater than 2, it is time to start worrying. Like all rules of thumb, this is only meant to be used as a rough guide. In reality, the worry threshold depends on things like sample size and the nature of the overdispersion. If in doubt, ask someone who knows about statistics to help you decide whether or not to worry. Alternatively, you could try out a different kind of model.

What are your options?

- One simple way to fix a model indicating overdispersion is by changing the family in **glm**() to a 'quasi' version of your distribution (e.g. `family = poisson` becomes `family = quasipoisson`). A 'quasi' model works exactly like the equivalent standard GLM model, but it goes one step further by estimating that dispersion index we mentioned above—it does it in a more clever way than we did. Once this number is known, R can adjust the *p*-values to account for it.
- Another simple way to fix an overdispersed Poisson model is to switch to the 'negative binomial' family. A negative binomial distribution can be thought of as a more flexible version of the Poisson distribution. The variance increases with the mean for a negative binomial distribution, but it does so in a less constrained way, i.e. the variance does not have to equal the mean.

Let's discuss briefly how these two options work in the real world.

A 'quasi-' model works exactly like the equivalent standard GLM model. If you build two versions of the Soay model—one using `family = poisson`, the other using `family = quasipoisson`—and compare the summary tables, you will see that the coefficient estimates are the same in both models (go on, try it). The only thing that should be different is 'the stats'. These are based on a method that accounts for the overdispersion (or underdispersion).

Simple, yes? Well, there is one more thing to be aware of. You should be careful when you use anova with a 'quasi' model, because you have to explicitly tell R to take account of the estimated dispersion. This is not hard to do. Instead of using **anova**(..., test = "Chisq"), you have to use **anova**(..., test = "F"). This tells R to use an *F*-ratio test instead of a likelihood ratio test, because this allows us to incorporate the dispersion estimate into the test. We won't say more, as there is too much statistical magic here to explain in a simple way.

Negative binomial GLMs are easy to work with. However, we *do not* use **glm**() to build these models; use **glm.nb**() from the **MASS** package instead. The **MASS** package is part of base R, so you won't need to install this; just use **MASS**. Other than this, there is very little new about using negative binomial GLMs. You don't need to worry about the family, because **glm.nb**() deals only with the negative binomial. There is a link function to worry about—the default is the natural log again, but you have a couple more options (take a look at `?glm.nb`) —and, of course, you have to use R's fantastic formula to define the model.

That's it for 'quasi' and negative binomial models. We could walk you through how to use them, but, really, you already have all the skills you need to do this on your own.

We'll finish off with a little warning, though. The 'quasi' and negative binomial tricks often work well when overdispersion is produced by missing variables. Sadly, they don't really fix overdispersion that is generated by non-independence. That kind of problem is better dealt with using more sophisticated models. This isn't the place to discuss them, but just in case you're the kind of person that *needs* to know, we'll tell you what the most common solution is called: a mixed model.

7.5.2 ZERO INFLATION

There's one very specific source of overdispersion in Poisson-like count data that's worth knowing about if you're a biologist: zero inflation. This happens when there are too many zeros relative to the number we expect for whatever distribution we're using. If your count is zero-inflated, you can often spot it with a bar chart of raw counts. If you see a spike at 0, that's probably (not always!) caused by zero inflation.

Biological counts are often zero-inflated. This often results from a binary phenomenon acting in combination with a Poisson process. For example, the number of fruits produced by an outcrossing plant depends first on whether it is ever visited by a pollinator, and then on the number of flowers visited. The number of zeros could be very high if there are few pollinators around because many plants are never visited, but for those plants that are visited, the number of seeds might be nicely Poisson distributed.

Zero inflation is best dealt with by using a new model. There are (at least) a couple of options available here:

- Option 1 is to use a *mixture model*. These work by modelling the data as a mixture of two distributions (sometimes statisticians do give things sensible names). They assume each observation in the data comes from one of these two distributions. We don't know which, but, fortunately, some clever statistical machinery can deal with this missing information.
- Option 2 is to use a *hurdle model*. A hurdle model has two parts. It has a binary part for the zeros (it asks whether the runner makes it over the hurdle—zero or not), and a Poisson part for the non-zero values (it asks how many steps they then take—a positive integer). The clever part is that the Poisson part is modified so that it only allows for positive values.

If you do run into a zero-inflation situation, R has you covered (of course). There are quite a few options available, but the most accessible is probably the **pcsl** package (available on CRAN). This provides a couple of functions—**zeroinfl()** and **hurdle()**—to model zero-inflated data using a mixture or hurdle model.

7.5.3 TRANSFORMATIONS AIN'T ALL BAD

You probably learned in your first stats course that you should always transform response data that are counts, and then use a general linear model. The log and square root transformations are the usual options for Poisson-like count data. Later on, if you did a more advanced course that included GLMs, you may have been told never to transform your count data. Instead, use a GLM, because... they're just better.

Who is right? We think they are both wrong. Sometimes transformations work, sometimes they don't. Sorry, we're pragmatists, which means we are OK with shades of grey. If you've designed a good experiment, you probably just want to know whether a treatment had an effect or not (*p*-values), and whether that effect is big or not (coefficients). If in doubt, try out a transformation, build the model, and check the diagnostics. If these look fine, it's probably OK to use the model. There *are* some advantages of using transformations:

- Transformations often work just fine when the data are 'far away from zero' (so there are no zeros) but don't span many orders of magnitude.
- Using a transformation is simple because there are no nasty link functions to worry about. But. . . you can also analyse fancy experiments like 'split-plot' designs fairly easily.
- You don't have to worry about overdispersion. The residual error term takes care of the overdispersion. This can be a big advantage.

So why don't we always use transformations? Sometimes you just can't find a good one:

- Transformations change two things at the same time. They alter the mean–variance relationship, and they change the 'shape' of the relationship between predictor variables and the response. A transformation might fix one, but break the other.
- Those zeros again. Transformations often fail when your count data contain zeros. You can't take the log of zero, so you have to use $\log(y + 1)$. This almost never produces a model with good diagnostics.
- The model you build may be difficult to interpret and use because it does not make predictions on the scale you originally measured things on.

If you aren't sure what to do, explore some options and follow the approach that we advocate in this book: *Plot -> Model -> Check Assumptions.*

7.6 Summary, and beyond simple Poisson regression

Well, that was fun, don't you think? You've just been introduced to the GLM. We've discussed data types that often necessitate (i.e. beg for) a GLM, including count data, binary data. We have deliberately introduced more theory here, as a way to understand general versus generalized linear models, and also to boost your confidence. Statistical modelling is not easy, but you have embarked or are just embarking on a career in research, and understanding deeply how to manage, communicate, and analyse your data is vital. You now know a great deal. Perhaps just enough to be *dangerous*.

7.6.1 THE LINK FUNCTION RULES

We hope, perhaps more than anything, that you now understand that when we specify a family to manage 'non-normal' data, we are making very specific assumptions about the nature of the variability in those data. This is very useful, because knowing these assumptions allows us to use all of the diagnostics and tools we are accustomed to from general linear models, even though things like odd mean–variance relationships are in play. The downside is that sometimes our data won't play ball with these assumptions. Overdispersion is a common problem in these cases.

We also introduce a link function into the mix whenever we build a GLM. The link function manages the fact that predictions for many types of data must be bounded. A model that predicts negative counts is not ideal, for example. We refer you again to Figure 7.4. If you understand that we have to move from the response scale to the scale of the link function to fit the model, and then back again to interpret and plot the results, you have won the game. Now you can read the hard books.

7.6.2 THE WORKFLOW STAYS THE SAME

In Chapters 5 and 6, we introduced a deliberate and recipe-like workflow of *Plot -> Model -> Check Assumptions -> Interpret -> Plot Again*. This

workflow does not change with a GLM. Some bits are harder. Making the
first and last figures can require more thinking. Luckily, when used ap-
propriately, interpreting the diagnostic plots remains as straightforward
as ever. Interpreting the **anova**() and **summary**() tables requires understanding a bit more about likelihood and deviance, but, at the same time,
the structure and methods for generating and interpreting these tables
remain the same. As the models you fit become more complicated (e.g.
interactions between variables), the **anova**() table is still telling you about
each part of the model (main effects and interactions), treatment contrasts are still the default in the **summary**() table, and alphabetical listing
of categorical-variable levels is still the norm.

7.6.3 BINOMIAL MODELS?

We have not covered binomial models, which, as we note, come in many
forms. We refer you now to more advanced books on the GLM, including a bible of the R community, *Modern Applied Statistics with S.*
(see Appendix 2). What you will find in your readings, and in your exploration of various web resources, is that the model formula works just as it
does in **lm**(), as with our Poisson GLM, but that there are several different
ways of coding the response variable—it can be binary, percentages with
weights, or a freaky two-column variable of success/failure from trials. Oh,
and the default link function is the logit link, with a couple more options
if you need them. We have to use a new link function because the goal of
a binomial GLM is to model probabilities of events, and these have to lie
between 0 and 1. Don't let these things scare you, though. You *are* more
than ready now to tackle more advanced readings and ideas.

8

Pimping Your Plots: Scales and Themes in *ggplot2*

8.1 What you already know about graphs

You have just emerged from the deepest of the statistical depths we will take you to: the generalized linear model. You are now ready for serious data analyses, handling the many sorts of experimental designs, sampling designs, and types of data we use and find ourselves collecting in biology.

Along the way, we've covered several tools for making figures that reflect these designs and data, mostly using *ggplot2*. We've also introduced a few 'customizations' that have helped emphasize features of our data, or change colours of points or fillings of bars. Overall, you should be relatively proficient with:

- using and even combining geom_point(), geom_line(), geom_boxplot(), geom_bar(), geom_histogram(), and geom_errorbar();
- using aes() with arguments colour = and fill = to assign colours or fills to points or bars based on categorical grouping variables;
- using size = and alpha = within aes() or geom_() to customize the point size and transparency of the points/bars/histograms;

Getting Started with R Second Edition. Andrew Beckerman, Dylan Childs, & Owen Petchey:
Oxford University Press (2017). © Andrew Beckerman, Dylan Childs, & Owen Petchey.
DOI 10.1093/oso/9780198787839.001.0001

- using `ymin =` and `ymax =`, generated via **dplyr**, within **aes()** in **geom_errorbar()**;
- using **scale_colour_manual()** and **scale_fill_manual()** to choose custom point and fill colours;
- using **theme_bw()** to customize overall features of the graph, including the background colour, etc.

That's quite a bit of plotting skill picked up 'along the way'. What we'd like to do now is give you a way to think about organizing this information in a way that allows you to continue to learn how to use and extend the features of **ggplot2**, and its powerful and productive interface with **dplyr**.

Before we dive into more **ggplot2** syntax, we'd like to emphasize and encourage you to use the Internet. Several online resources we mentioned in the earliest chapters are worthy of mentioning again. Foremost are the web pages and cheat sheets for **dplyr** and **ggplot2**. The second is Stack Overflow and the 'r' channel found currently at `http://stackoverflow.com/tags/r/info`. Using *natural language* queries in Google or your search engine of choice may be surprisingly effective. For example, do not be afraid to type *'How do I add emojis to a graph in ggplot2?'*.

OK. Let's move on to some detail. We have no intention of covering the vast array of detail and customization that can be achieved in **ggplot2**. But what we can do is give you a taster of what you can do. To do this, we'll step back to the dataset we started learning R with, and started making graphs with. The `compensation.csv` data, where cows graze in lush fields under orchards, helping achieve paradise, happy dirt, and juicy apples.

8.2 Preparation

You may again want to start another new script. Make some annotation, save it as something constructive like `ggplot2_custom_recipes.R`, and make sure you have **ggplot2** loaded as a library (e.g. using **library**(ggplot2)). Furthermore, go and get a new package from CRAN, called **gridExtra**, download it, and then make it available with **library**(gridExtra). Finally, grab the compensation data again, and we'll begin.

8.2.1 DID YOU KNOW ... ?

We are going to start with the scatterplot of Fruit versus Root, as we did in Chapter 4. We are also going to build a box-and-whisker plot, treating the Grazing variable as the two-level categorical variable that it is. In addition to this, we are going to assign the graphs to objects eg_scatter and eg_box. This allows us to easily use the graphs again. Here are the two graphs to make:

```
# BASE scatterplot
eg_scatter <-
  ggplot(data = compensation, aes(x = Root, y = Fruit)) +
  geom_point()

# BASE box-and-whiskers plot
eg_box <-
  ggplot(data = compensation, aes(x = Grazing, y = Fruit)) +
  geom_boxplot()
```

Now we can use the figure and even add to it. For example, let's apply theme_bw() to the eg_scatter figure (Figure 8.1):

```
eg_scatter + theme_bw()
```

Now, we also had you download a new package, *gridExtra*. *gridExtra* is a helper package for, amongst other things, placing more than one *gg-plot2* figure on the same page. Yes. Not only can you make figures with many facets/panels using *ggplot2*, but you can also then place many of these many-faceted figures onto the same page. Sweet (Figure 8.2):

```
grid.arrange(eg_scatter, eg_box, nrow = 1)
```

The arguments provided to **grid.arrange**() are simply a list of the graphs, separated by commas, and an arrangement specified by nrow = or ncol = or both! Take a look at the help file and examples. . . there is quite a bit of fun to have here.

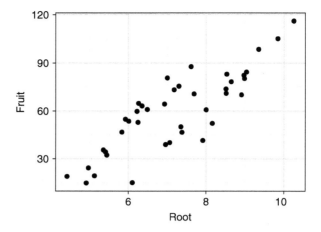

Figure 8.1   We can make a plot, assign it to an object, and then add layers to this.

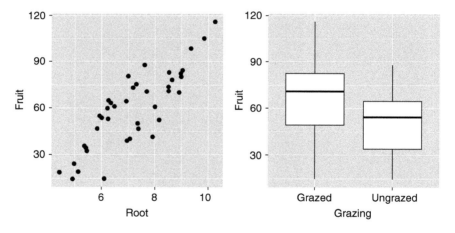

Figure 8.2   The base scatterplot and a boxplot, using the compensation data and organized onto one page using grid.arrange from ***gridExtra***.

8.3 What you may want to customize

When we teach R, we always get to a point where we ask the class: What do you want to change about the graphs!? It's rather fun, and produces a consistent list of things to change. Mostly, it is these:

- axis labels, with maths symbols, rotation, and colours;
- axis ranges and tick mark locations;
- the grey background *and* the gridlines;
- the boxes and title in the key;
- text annotation inside the plot.

Making these things happen requires being aware of what are, broadly speaking, two routes to customizing a **ggplot()** figure. The first is via a scale_() function of some kind. The second is via the **theme()**. The theme essentially determines how the parts of a graph that are *not* directly determined by the mappings in **aes()** are displayed. These are things like the gridlines and text formatting.

The scale_() functions, in contrast, are deeply tied to the *x*- and *y*-axis variables, which in *ggplot2*-land are defined by the *aesthetic* mappings defined by **aes()**. You may not have realized it, but every time you establish a mapping between a variable in your data and an aesthetic, you define something called a *scale* in **ggplot()**. For example, in our scatterplot, we mapped Root to the *x*-axis. In the process, *ggplot2* has captured the range of the rootstock data and defined the breakpoints where to put tick marks, amongst other things.

If you managed to follow that, you might realize now that there are pieces of the graph linked directly to a particular variable, like ranges and breakpoints; this is the remit of the scale_() functions. There are also pieces of the graph, like the presence and absence of gridlines, the rotation and colour of text on the graph, etc., not tied to an aesthetic mapping; these are the remit of **theme()**. Finally, there are the words and word choices we might use to name axes, or words to put anywhere on the graph (annotation). These annotations have their own special functions. . . and we are going to start with these.

8.4 Axis labels, axis limits, and annotation

We can change the axis titles with **xlab()** and **ylab()**, or their parent and guardian function, **labs()**. It is critical to note that these functions only tell

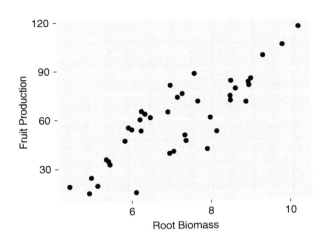

Figure 8.3   Using **xlab()** and **ylab()** to customize the words on the axes.

R *what* to show, for example the words to use. Be patient. . . we'll come to colours, rotations, and more in just a moment.

Here is how you can use **xlab()** and **ylab()** (Figure 8.3):

```
eg_scatter + xlab("Root Biomass") + ylab("Fruit Production")
```

If you really must include a title, use **ggtitle()**. . . though we ask those of you in academia when and where was it you've ever seen a title on a graph in a journal? But we digress:

```
eg_scatter + ggtitle("My SUPERB title")
```

As we note above, it is possible to combine these in one go with **labs()**:

```
eg_scatter + labs(title = "My useless title",
            x = "Root Biomass", y = "Fruit Production")
```

It is also convenient to change the range of the *x*- and *y*-axes. There is a more sophisticated method for this, allowing changes of the range *and* the breakpoints for tick marks, but that is the remit of the **scale_()** functions and we'll come to that in a moment. The convenience functions are the same as they are in base graphics, if you've used them. They are **xlim()** and **ylim()**:

```
eg_scatter + xlim(0, 20) + ylim(0, 140)
```

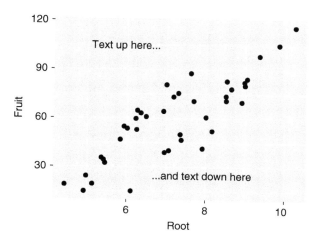

Figure 8.4   Adding text annotation inside the figure.

8.4.1 PUTTING CUSTOM TEXT INSIDE THE FIGURE

The **annotate()** function allows you to place custom text inside the graph/figure. This layer requires specifying what kind of annotation to use, for example 'text', where to put the annotation using the coordinate system of the graph, and what to write—the label. Here we provide an example of placing two pieces of text, the first at $x = 6$ and $y = 105$ and the second at $x = 8$ and $y = 25$. Of course you can do each piece on its own, but this shows you how easy it is to provide more than one piece of information to a function in *ggplot2* (Figure 8.4):

```
eg_scatter +
    annotate("text", x = c(6,8), y = c(105, 25),
             label = c("Text up here...","...and text down here"))
```

8.5 Scales

The **scale_()** functions, as noted above, are deeply tied to the variables we are plotting. *ggplot2*, like all good plotting packages, sets some defaults when you decide to make a graph. It grabs the variables mapped to each

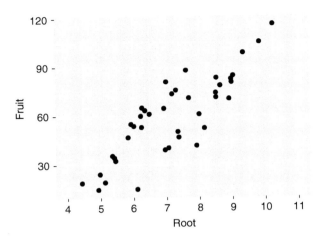

Figure 8.5   Changing the limits and breakpoints of a continuous x-axis.

axis and specifies several default features of the axes. We can of course
change these. Here we adjust the range of the x-axis using the `limits =`
argument, and the frequency and location of the tick marks by specifying
the integer values where we want them with the `breaks =` argument.
We extend the range of the x-axis to go from 4 to 11, and place tick marks
in steps of 1 between 4 and 11 (Figure 8.5):

```
eg_scatter + scale_x_continuous(limits = c(4, 11), breaks = 4:11)
```

Another feature of the **scale_()** class of adjustment is associated with
the colours and fills of the geometric objects we put on a graph. If you recall from Chapter 4, we learned how to use the `colour =` argument in
aes() to allocate different colours to points based on their membership of
a group. We extended this functionality by customizing the colours allocated to each group as follows, using the **scale_colour_manual()** layering.
We modify the *eg_scatter* example here to have a `colour =` argument
in the **aes()** portion of **ggplot**, and provide a custom set of colour values, brown and green, which get allocated respectively to Grazed and
Ungrazed, the alphabetical ordering of the levels of the Grazing factor
(Figure 8.6):

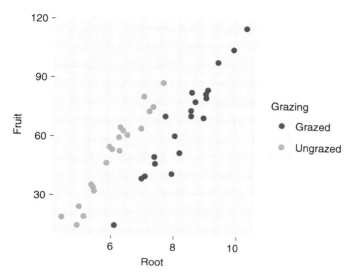

Figure 8.6 Changing the colours associated with levels of a grouping variable.

```
ggplot(data = compensation, aes(x = Root, y = Fruit, colour = Grazing)) +
  geom_point() +
  scale_colour_manual(values = c(Grazed = "brown", Ungrazed =
                      "green"))
```

You can leverage the functionality of **scales_()** to also transform an axis of a plot. For example, we may wish to log-transform the *y*-axis of a plot to manage some non-linearity or emphasize the extent of variation. We can make this happen 'on-the-fly' with *ggplot2*. We use the *boxplot* we created and generate a log-*y* axis. We use the `trans` = argument in **scale_y_continuous()**. `trans` =, as specified in the help file for **scale_y_continuous()**, can allocate many transformations to either or both of the *x*- and *y*-axes (Figure 8.7):

```
eg_box +
  scale_y_continuous(breaks = seq(from = 10, to = 150, by = 20),
                     trans = "log10")
```

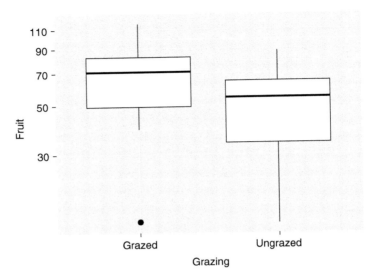

Figure 8.7 Transforming an axis with log10 in the graph, and specifying tick locations using breaks.

We draw your attention to how we have used **seq**() to generate the vector of breakpoints at which we wanted the ticks, and then asked for the log10 transformation. The **scale_**() help files are replete with several very informative examples, all of which work by cutting and pasting. You'll get a great deal out of looking through these now with the basic understanding we've given you.

8.6 The theme

Our final instalment of *ggplot2* customization familiarization funifications is associated with the **theme**() function, a very powerful framework for adjusting all the non-aesthetic pieces of a graph, and creating both beautiful and downright ridiculous figures. We show you here a few examples of customization that should set you up for using the excellent help file for **theme**() and stimulate your creative juices. We also note that several clever folk out there have made some very good customized themes that may actually suit your needs. You can find several within *ggplot2*, such as **theme_bw()**, by investigating the help file for **ggtheme()**. The package *ggthemes*, which has several custom themes that emulate base R graphics, the *Economist*, and even (we quote) 'the classic ugly gray charts in Excel'.

8.6.1 SOME **theme()** syntax about the panels and gridlines

Let's look at the syntax of a few theme adjustments that you may want to make. For example, let's see how to get rid of the grey background and minor gridlines, but generate light blue major gridlines. You may never want to do this, but it shows you the facility associated with the *panel* group of theme elements (Figure 8.8):

```
eg_scatter +
  theme(
    panel.background = element_rect(fill = NA, colour = "black"),
    panel.grid.minor = element_blank(),
    panel.grid.major = element_line(colour = "lightblue")
  )
```

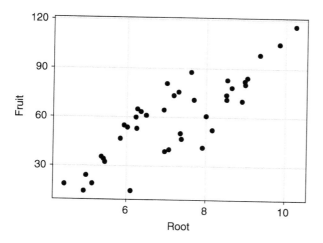

Figure 8.8 Adjusting the panel background and the gridlines using the **theme()** layer in *ggplot2*.

Let's review the several features of this set of customizations:

- The *panel* group of theme elements have some logical names, like 'background' and 'grid', corresponding to clear features in a figure.
- element_() specifies arguments for the *panel* group geometric components, such as 'rect' for 'rectangle' or 'line' for 'lines'.
- Each of these element_() arguments can be customized with now familiar (we hope) arguments such as `fill` = and `colour` =, as we have used via panel.background = **element_rect**(fill = NA, colour = "black").
- There is a sledgehammer of a tool, **element_blank()**, that shuts down all rendering of a **panel.()** component, as we have used for panel.grid.minor = **element_blank()**.

8.6.2 SOME **theme()** SYNTAX ABOUT AXIS TITLES AND TICK MARKS

If you've had a look at the help files for built-in themes like **theme_bw()**, you will have noticed that there is an argument in all of these custom themes for `base_size=`. This simple tool for increasing font sizes is currently not available via simply using **theme()**. If you want to change how the axis tick mark labels and axis labels are formatted, you must manipulate the theme by setting attributes of **axis.title()** and **axis.text()** components, which do indeed come in *x*- and *y*-flavours.

For example, here is how we can adjust the colour and size of the *x*-axis title, and the angle of the *x*-axis tick labels. We do this with the *eg_box* example, where the rotation can be super-handy for text-based labels (Figure 8.9):

```
eg_box +
  theme(
    axis.title.x = element_text(colour = "cornflowerblue",
       size =rel(2)),
    axis.text.x  = element_text(angle = 45, size = 13, vjust = 0.5))
```

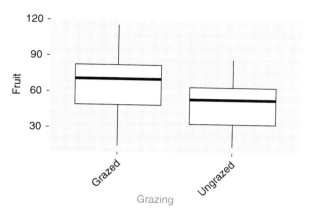

Figure 8.9   Modifications of the axis and tick labels are done with the 'axis.' class of **theme**() adjustors.

Once again, we can see that this **axis.**() class of attributes centres on manipulating text elements (**element_text**()), and these take several familiar arguments, and some that may not be so familiar. For example, `size =` can take an absolute size (e.g. 13) or a relative increase over the default, using for example `size = rel(2)`.

We further note the `vjust =` argument, which takes values between 0 and 1, to 'vertically adjust' the labels, which is often necessary when you rotate words.

8.6.3 DISCRETE-AXIS CUSTOMIZATIONS

The last thing we'll cover about axes involves some specific alterations to discrete axes. Discrete axes, as in our boxplot example, delineate different groups of data according to the values of a categorical variable. These variables are represented by a factor or character vector in R, which typically has a small number of unique values. These values are used to name each group if we do not provide an alternative. We use the scale_x_discrete() or scale_y_discrete() function to customize the labelling of groups.

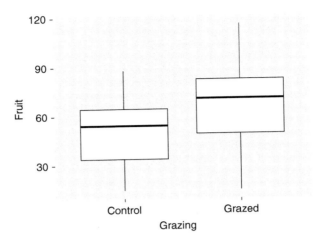

Figure 8.10 Changing the labels of discrete axes from the default use of the levels in the data frame to other names.

For example, we can alter the names of the levels on our graphs via (Figure 8.10)

```
eg_box + scale_x_discrete(limits = c("Ungrazed", "Grazed"),
                          labels = c("Control", "Grazed"))
```

8.6.4 SOME **theme()** SYNTAX ABOUT CUSTOMIZING LEGENDS/KEYS

It is our experience that minutiae can often generate anxiety, frustration, and the wasting of precious hours, days, weeks, and even years. One of such minutiae is customization of the key or legend that ***ggplot2*** produces. A quick look at the help file can save hours.

Perhaps you don't like the boxes, preferring a 'clean' key (Figure 8.11):

```
ggplot(compensation, aes(x = Root, y = Fruit, colour = Grazing)) +
  geom_point() +
  theme(legend.key = element_rect(fill = NA))
```

Sometimes you don't want a key at all. We can invoke the magic sledge-hammer **element_blank()** to whip it, whip it good. Interestingly, we do this using the **legend.position()** attribute (Figure 8.12):

```
ggplot(compensation, aes(x = Root, y = Fruit, colour = Grazing)) +
  geom_point() +
  theme(legend.position = "none")
```

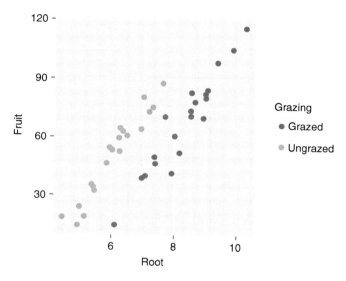

Figure 8.11 We can get rid of the bounding grey box in the legend with the legend.key feature.

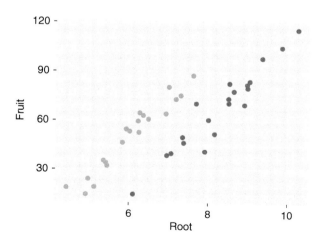

Figure 8.12 We can get rid of the legend entirely too.

Interestingly, again, this gives us the deeply satisfying insight that we can move the legend anywhere we want. We can. Check out. . . yes. . . the examples in the **theme()** help file.

8.7 Summing up

We hope this gives you some insight into how to use the functionality of scale_() functions and **theme()** attributes in ***ggplot2***. We've been relatively modest in the pimping—one can change just about everything associated with both the mapping of your variables to structural features of the figure, and the atttributes of the figure that are independent of the variables. Our intention is to have provided a close-up look at the syntax associated with using scale_() functions and **theme()**. You are ready now to delve deeply into the help file for **theme()**. Perhaps you'll even craft your own! Many people do. And don't forget the help file for **ggtheme()** or the package ***ggthemes***.

Go forth and make figures.

Closing Remarks: Final Comments and Encouragement

So, you really do know how to use R. You have acquired skills that allow you to import data, summarize and explore it, and plot it in a wide range of formats. You have acquired skills as well that allow you to use R to do basic statistics. Throughout, we have ensured that you understand the value of writing down and saving the instructions you want R to carry out in a script. As a result, you end up with a *permanent, repeatable, annotated, shareable archive of your analysis*. Your entire analysis, from transferring your data from your notebook to making figures and performing analyses, is all in one, secure, repeatable, annotated place.

We also have introduced you to a very fundamental rule for data analysis. *Never* start an analysis with statistical analysis. *Always* start an analysis with a picture. Why? If you have done a replicated experiment, conducted a well-organized, stratified sampling programme, or generated data using a model, it is highly likely that some theoretical relationship underpinned your research effort. You have in your head an *expected* pattern in your data. Make the picture that should tell you the answer—make

Getting Started with R Second Edition. Andrew Beckerman, Dylan Childs, & Owen Petchey:
Oxford University Press (2017). © Andrew Beckerman, Dylan Childs, & Owen Petchey.
DOI 10.1093/oso/9780198787839.001.0001

the axes correspond to the theory. If you can do this, *and* see the pattern you *expected*, you are in great shape. You will know the answer!

Throughout the book, we emphasized a workflow centred on graphing and statistical analysis. Here we expand this workflow to 17(!) steps. The additional steps here include such things as snacks and organization.

1. On your computer, create a folder for your project.

2. Enter your data into a spreadsheet (e.g. Excel), with observations in rows and variables in columns. Save it to your new folder.

3. Print a hard copy of the spreadsheet and check it against the original data sheets. Correct any errors in the spreadsheet.

4. Save the spreadsheet in comma-separated values (`.csv`) format.

5. Protect the spreadsheet and `.csv` file by making them read-only files.

6. Take a coffee, tea, or other beverage break.

7. Start up RStudio, open a new script file, and save it to the folder you created in step 1. Write your instructions for R in this script file, and save it regularly (really!). This script file and your data file are all that you need to keep safe (but you need to keep them really safe!).

8. Always put comments in your script.

9. Import your data into R. Check that the numbers of rows and variables are correct. Check that variable types are correct. Check that the number of levels of categorical variables is correct. Check that the range and distribution of numeric variables are sensible. Check for any missing values and make sure you know where they are. Check how many data points per treatment combination. Check everything!

10. Don't forget to put comments in your script.

11. Have a snack break.

12. Explore your data, primarily by using R's awesome plotting functions. *dplyr* and *ggplot2* are particularly useful at this stage. Guess the answer that your statistical tests should give you. Make some

other graphs to further investigate your guesses. Make some more figures, just to be sure.

13. Use a statistical test to check if your guess is right or wrong. If it's right, rejoice and publish. If it's wrong, figure out why.

14. Don't forget to put comments in your script.

15. Communicate your findings to your colleagues, in an email, report, manuscript, poster, web page, etc.

16. Go back and organize the files in your folder and the script you've written. Leave it all in a state that you'll be happy to come back to in six months' time—that's a long time from now, so take your time doing this.

17. Retire to your preferred socializing establishment. Try not to bore your non-R friends by talking about R all night.

If you stick to this basic workflow you will have a very solid and efficient foundation for using R. Yes, there are many ways you could go about doing what we described above. However, why not just use one way and keep your R-life simple? Make all of your scripts and analyses cover each of these steps. Make sure you annotate your script so that each of these sections is clear. If you do all of this, you will be an efficient and happy R user. And you, your friends and colleagues, and editors will appreciate all of the effort when you need to revisit, share, or publish your analysis.

Happy R-ing.

General Appendices

Several chapters possess specific appendices. Here we provide an additional three that are not chapter specific, and offer fuRther oppoRtunities for leaRning and joculaRities.

Appendix 1   Data Sources

Table A.1   Sources of datasets used in this book.

Dataset	Original source
compensation.csv	Crawley (2012), ipomopsis.txt dataset.
gardenozone.csv	Inspiration from Crawley (2012).
	Dataset used in this book synthesized by the authors.
plant.growth.rate.csv	Dataset synthesized by the authors.
daphniagrowth.csv	Dataset synthesized by the authors.
growth.csv	Crawley (2012), growth.txt dataset.
soaysheepfitness.csv	Dataset synthesized by the authors.
limpet.csv	Quinn & Keough (2002).
ladybirds_morph_colour.csv	Dataset synthesized by Phil Warren.
nasty format.csv	Dataset provided by OP.

Appendix 2   Further Reading

A.2.1 Online

- `https://journal.r-project.org` —*The R Journal.*
- `https://cran.r-project.org` —see the Documentation section, where there are manuals and guides in many languages.
- `https://www.rstudio.com/resources/cheatsheets/` —some beautiful cheat sheets for things such as data visualization, data wrangling, base R, advanced R, and R Markdown.
- `http://ggplot2.org` —a comprehensive guide to making beautiful graphics with ***ggplot2***.
- `http://www.statmethods.net` —the Quick-R website. Lots of useful examples and lessons.
- `http://stackoverflow.com` —lots of questions and answers about R here. Searches in Google will often end up here.
- `http://www.r-bloggers.com` —aggregates many R-related blogs.
- `https://www.datacamp.com` —online R courses. Also check out courses on edX, coursera, and other online learning platforms.

A.2.2 Print

- H. Wickham, *ggplot2: Elegant Graphics for Data Analysis*, Springer, 2009.
- M.J. Crawley, *Statistics: An Introduction using R*, Wiley, 2005.
- M.J. Crawley, *The R Book*, Wiley, 2012.
- A. Hector, *The New Statistics with R: An Introduction for Biologists*, Oxford University Press, 2015.
- P. Dalgaard, *Introductory Statistics with R*, Springer, 2008.

- J. Maindonald and W.J. Braun, *Data Analysis and Graphics Using R: An Example-Based Approach*, 2nd edn, Cambridge University Press, 2010.
- J. Fox, *Applied Regression Analysis and Generalized Linear Models*, 3rd edn, Sage, 2015.
- J. Fox and S. Weisberg, *An R Companion to Applied Regression*, 2nd edn, Sage, 2011.
- J.J. Faraway, *Linear Models with R*, 2nd edn, CRC Press, 2014.
- J.J. Faraway, *Extending the Linear Model with R*, CRC Press, 2005.
- W.N. Venables and B.D. Ripley, *Modern Applied Statistics with S*, Springer, 2002.
- J. Adler, *R in a Nutshell*, O'Reilly, 2012.
- G.P. Quinn and M.J. Keough, *Experimental Design and Data Analysis for Biologists*, Cambridge University Press, 2002.

Appendix 3   R Markdown

When you communicate your results, you may find yourself copying and pasting output from the Console (e.g. an ANOVA table) to a Word document or PowerPoint presentation. And saving a figure from R, and pasting it into the same documents. This works just fine. Though if your dataset changes, you have to copy and paste again. No big deal.

Another option is R Markdown—a special type of script that contains a mix of plain text, formatted text, and R code, and when 'Run' produces a document (HTML, Word, or PDF format) or presentation (HTML or PDF format). The plain text appears as plain text, the R code appears as R code, and, most important, the result of the R code is inserted into the document. If the R code produces a graph, then the code and graph, or just the code, or just the graph can get inserted into the document. If the R code returns a summary from a linear model, then the model and summary, or just the model, or just the summary can get inserted into the document.

Best of all, it is very, very easy to use, in large part thanks to RStudio. When you ask RStudio to make a new file, one of the options is *R Markdown*. Click on this, and in the dialogue box choose what type of document you want to make. This opens a new script file... but wait, it already contains some script. This pre-supplied script talks you through the rest of the process, including how to press the *Knit* button to make the final document.

R Markdown is fantastic. Explore it, use it, promote it. If you need even more, check out Sweave, and also Shiny Apps.

Index